Biology and Ethics
by
Rodrigo Fernós

September 2018
San Juan, Puerto Rico

"Biology and Ethics," by Rodrigo Fernós. ISBN 978-1-949756-09-8.

Published 2018 by Virtualbookworm.com Publishing Inc., P.O. Box 9949, College Station, TX 77842, US. ©2018, Rodrigo Fernós.

Preface

The study of biology and ethics, as many other multidisciplinary endeavors, is fraught with complexities. One the one hand, representatives from the Catholic Church in Puerto Rico have taken over the field of 'bioethics', stripping it entirely of its historical character. In 2016 I met Jesuit priest Jorge José Ferrer in charge of a new center for bioethics, who sincerely told me that bioethics was a purely philosophical endeavor.[1] An article on the topic by Leonides Santos y Vargas, former director of the organization, took a similar position, in which bioethics is defined as a study only of the misuses of genetics, which by definition begins only after World War II with the discovery of recombinant genetics allowing for the direct manipulation of this fragile-yet-persistent material.[2] While Catholic thinkers would be naturally hesitant to include the study of ethics from a biological point of view, whose early efforts suffered many theoretical problems and horrific implications for public policy, one cannot help but notice that all human actions are measured and judged in a historical context. Just as in the physicists universe, there is no static framework from which to view human actions—except a relativistic one provided by Darwinian evolution.

Tragically and oddly, the creation of the organization led by Ferrer was preceded by the destruction of the only organization in Puerto Rico dedicated to the history and philosophy of science; it

---

[1] This view was repeated in his article, *"La bioética como quehacer filosófico"* (2009). While he recognizes the importance of transdisciplinary approaches and the delimited historical character of bioethics, he builds a strawman argument. His most forthright definition of bioethics is as follows: *"la busqueda de soluciones morales justificadas en el contexto de las sociedades pluralistas y complejas que habitamos."* (p. 39). In spite of his explicit intentions, in practice his analysis is ahistorical, as we will see in the book.

[2] Leonides Santos y Vargas, *"Bioetica y Sociedad"* (1990).

is upon the bones of the *Centro para la Filosofía e Historia de la Ciencia y la Tecnología* at the University of Puerto Rico in Mayagüez that the *Instituto de Bioética* at the UPR (Recinto de Ciencias Médicas) was founded.[3] The history of science institute was originally headed by Elena Lugo, whose strong ties to the Catholic Church might have led her at one point to restructure the organization's role and purpose. Lugo turned the history of science center into the *Centro de la Ética para las Profesiones*, becoming a precursor to the current organization. It appears that by adopting the name of a famous Puerto Rican thinker who lived in Chile most of his life, Eugenio María de Hostos, the serial institutional changes would be legitimized. Puerto Rico, as suggested by Immanuel Wallerstein, appears to sit on the semipheriphery, a site in which global battles are fought by heavyweight contenders, in this particular case between the Catholic Church and modern science.[4]

On the other hand, the most sophisticated efforts to view human behavior from an evolutionary point of view, by definition an extension of the biological paradigm, are also attacked by some biologists themselves. In his 'last lecture' on June 8, 2018 at the University of Puerto Rico (Rio Piedras), the retiring biologist Nicholas Brokaw provided such a straw man depiction of sociobiology, that it was unclear whether Prof. Brokaw either did not understand the field or whether he was simply mocking it. Brokaw, whom had previously worked at Barro Colorado Island in Panama, sarcastically noted that he was a biologist because of his 'hunter gatherer' heritage some 120,00 years ago.[5] Even though Brokaw clearly identified himself with the naturalist tradition in biology, he is apparently unwilling to recognize the revolutionary theory at his doorsteps as genetics is driving biology into a new scientific paradigm and modus operandi

---

[3] Marlen Oliver Vaïzquez, ed., *Ensayos en bioética: una perspectiva puertorriqueña* (2013), p. 32.

[4] This should not be taken as an attack on the Catholic Church, as it serves as a strong counter-weight to the powerful pharmaceutical companies dominating the island's political undercurrents.

[5] Nicholas Brokaw, "The Professor as Hunter-Gatherer", talk at the College of Natural Sciences, UPR (Rio Piedras), June 8, 2018.

which many in the field are uncomfortable with.[6] Biology is increasingly defined by complex algorithms driven by costly supercomputers, creating a new dystopia of 'haves' and 'have nots' in the field. This recent redefinition of role and purpose ultimately ignores its fundamental subject: the natural living world. Brokaw's comments might simply be reflecting this preoccupation.

We should not be so rash to reject 'transdisciplinary' efforts as sociobiology. As John Dewey so poignantly noted in 1908 with regard to the social implications of Darwinian evolutionary theory, it will alter Western Civilization's legal, moral and religious landscape. We should not allow disciplinary restrictions, so pervasive in today's specialized modern world, to stand in the way of scientific advancement.

Just as revolutionary theories do not necessarily discard prior theories but more accurately appropriate these within their findings, revolutionary theories in ethics should reinforce prior theoretical efforts in the historical record. As we will see, while the United States founding fathers, specifically Benjamin Franklin and Thomas Jefferson, lived a full two centuries prior to the emergence of genetics, and would have had no idea what sociobiology meant, many of the underlying principles of their work can be defined as "sociobiological" in character. As in physics, new biological theories of ethics should have some sort of historical echo with prior ideas, given that the evolutionary base of human behavior has changed relatively little in the last 200,000 years.

---

[6] Biology has been roughly divided during the 20th century into the 'geneticists' versus 'naturalists' tradition.

Acknowledgements

I would like to thank José García at the Biology Department (UPR Rio Piedras) for making this book a possibility. A special thanks is extended to both Nivea Santiago and her colleagues at the Natural Sciences Library (UPR, Rio Piedras) for their diligent efforts following Hurricane Maria on September 19, 2018. A word of special appreciation is also due to Gabriel E. De Jesus Astacio, Sylvia S. Eliza Pérez, Natalia Figueroa Ramos, Alejandro García Lugo, Cristina Hernández Riesco, Marelys E. Martínez Rodriguez, and María del Pilar Ortiz Fullana for their insightful comments on the topic. Finally, but not least important, I would like to extend a word of appreciation to Yarelis Torres Vázquez, who has assisted this endeavor in more ways than I could have imagined.

Table of Contents

Introduction

*Y mientras discuten las ciencias y la religión, el
hombre, pobre ciego vaga por el mundo guiado por un
lazarillo, la razón: y sometido a leyes eternas e
inmutables, que apenas vislumbra, vive y muere sin
saber de donde viene, ni a donde vá.*
- R. Gandía Córdova, 1902

It could be argued that there are two visions to bioethics. The
first is by Aristotle, who claimed that biology was the best
science, whose objects could be touched, allowing the scientist
direct and personal interaction with these. Biological objects
could be contrasted with those of astronomy. While stellar
phenomena as the planets or the Milky Way were aesthetically
pleasing, they were unreachable and elusive, and did not endow
men with a moral character. The second view is that of Leonardo
da Vinci's, who believed that humanity would end up destroying
the natural world. To him, there was an utter lack of morality in
human behavior, perhaps the outcome of his personal relationship
and employment to the murderous Borgias in Rome.

Which view is the valid one? These two view are obviously
two extremes, but are they mutually exclusive to one another?

There has been, and still exist, a great deal of abuse of
medicine and biology in history, the most notable being that of
Nazi Germany during World War II. All sorts of noxious
experiments were conducted, the Joseph Mengele twin studies
being one of the most notorious. However Mengele was not the
only one. The Allies also conducted their own, now well
documented, cases of human radiation experiments after the war.
The subjected patients were never informed that were being
radiated, turning these into non-consenting human guinea pigs
after the Nuremberg Trial. Cases as these are particularly noxious

given that the physicians who participated in them had all pleaded obedience to the Hippocratic oath, of inducing no harm willfully. The oath became a piece of toilet paper for some. At first glance, it would appear that da Vinci's view is the correct one.

While we will look at cases of torture and abuse as these, these will not be exclusively studied; otherwise, the book would end up as a rather somber and morose one. A study on torture would be quite boring from an academic point of view, and perhaps make psychopaths of some readers. It is much more interesting to ask questions pertaining to the theoretical underpinnings of human behavior, and analyze how these theories have changed over time. The book will thus be "Janus faced", looking to the past and to the future.

By the 'past' we are referring to the *long duree* of history and the biological underpinnings of ethics and human behavior—or, in other words, to evolutionary psychology or 'sociobiology' as it was originally called. Ethics is a uniquely human concern in that it has traditionally been conceived that animals do not have ethics, generally speaking, because their route of action is determined by instinct—or so the story goes. In order to have an ethical dilemma, one has to have freedom of choice, whereby the individual is presented with a set conflicting values from which he or she must decide and act. Do you sacrifice yourself for the collective, for example? Such an action is generally regarded as the key definition of altruistic behavior or ethical behavior properly speaking.

The notion of a biological basis of ethics first emerged in Darwin's *Descent of Man* (1871), who speculated on the issue. This was extended in a second movement by Herbert Spencer in the Darwinian century, which was itself modified by the rather noxious eugenics movement in the United States. However, during the 1960s and 1970s, there emerged a third attempt, that was much more rigorous and revolutionary: sociobiology. It participants included William Hamilton, George Williams and Robert Trivers; E. O. Wilson was the most well known for his book *Sociobiology* (1975). A truly a new way of looking at the world emerged from their work. As Theodosius Dobzhansky noted, one cannot understand biology without evolution; similarly, one cannot hope to understand human behavior without looking at its evolutionary biological underpinnings, which are more complex than is typically suspected. One of the best exposition of the field can perhaps be found in Robert Wright's *The Moral Animal* (1994).[7] Here the definition of the past is alluded to as "deep time", or the era between the written historical record (6,000 years or so) and the origin of *homo sapiens sapiens*—a 200,000 long year history which profoundly shaped the human psyche.[8]

The relative 'historical present', referring to the atrocities which have occurred during the 20th century, will also be analyzed. While such examples may shock sensitive souls, if we are not aware of extremes of human behavior, we will lack realism as thinkers, in that we will within a fictional universe of how biologists and doctors have actually behaved.

One of the problems with the notion of philosophy as the only valid approach to ethics is simply that all ethical dilemmas do not occur in a historical vacuum; prominent cases create a referent context for the evaluation of actions and expectations. The book will look at a broad range of cases, from the Tuskegee experiment with syphilis on African Americans to the radiation experiments during and After WWII alluded previously. The atrocities suffered by both soldiers and prisoners are typified by many common features, including the fact that patients were rarely informed that they were being treated with injurious substances. We will also look at the use of medicine for the purposes of political repression in Puerto Rico as the case cases of Cornelius Rhoads and Albizu Campos. There is a pattern behind all of these cases. It is often the weak and the vulnerable that are subjected to medical torture; they can easily be forgotten and discarded.[9]

---

[7] A good history of the biological underpinnings can be found in Paul Faber's *Temptations of Evolutionary Ethics* (1994). A more concise description can be found in Matt Ridley's *Origins of Virtue* (1998).

[8] The story reaches even deeper into the past than we can imagine.

[9] They have been aptly described in Suzan Lederer's *Subjected to Science.*

Finally, we will look to the future, specifically the impact of technological changes on ethical decision-making. Biological and medical research does not occur in a vacuum, as these are influenced by their institutional arrangements and the technologies available at the time of experimentation. The 20th century saw the rise of the corporation: profit seeking entities with vast financial resources available to them, but blindingly guided by the dictates of the market, which in turn tend to have a very corrosive impact on human behavior.[10]

Generally speaking, however, the nature of biology has changed over time, and in this sense the historical evolution of biology has been a determining factor in bioethics. The ethical dilemmas of biology have varied over the course of time, emerging out of both its theoretical and practical implications. We may note that natural history was distinctly non-experimental. This did not did not mean, however, that there were no discoveries proper—as shown by the Comte de Buffon, whose mathematical correlations and hypothesis regarding the origins of the Earth helped pave the way for our modern Darwinian view. Buffon's views went directly against French Catholic positions of the day.[11] How did experimentalism come to define modern biology and its ethical dilemmas? During the Illustration Diderot believed France to have been on the cusp of a biological revolution: life molecules during the Enlightenment obviously were never discovered.

This study will place the changing notions of ethics firmly within the general historical evolution of biology. Biology today might be defined as being in a 'postmodern phase' where computers have become deeply integrated into biological research. As physics, biology today is dominated by a strong statistical component. Can biology ultimately be reduced to computer modeling as well? Consequently, if all of biology is

---

[10] The best case studies are perhaps those by Marcia Angell's *Truth about Drug Companies* (2004) and Don K. Price's *The Scientific Estate* (1965), both of which present two distinct views on the issue.

[11] Georges Cuvier, a leading scientists of Museum of Natural History, studied comparative anatomy, and provided a key to understanding the structure of creation.

reduced to statistics, will experimentalism still be necessary? This view was incidentally held by Descartes and Galileo, who presumed the book of nature was written in the language of mathematics, and hence we only needed to study geometry to understand all of nature. The 'experimentation' done by Galileo was undertaken for rhetorical purposes only, and allegedly not for discovery proper. Will animal experimentation similarly become obsolete and create a situation where breaches of ethical norms, at least with regard to animal experimentation, no longer exist?

Let us explore.

Part I:
Early Theories of Human Nature

Darwin and the Moral Sense

The *Origin of Species* (1859) is perhaps the most widely sold and read scientific book in history. There are so many translations and editions, that it is actually hard to calculate their total number. The reason for the attention to the work is easy to identify. Its scientific argument is understandable: variation and natural selection lead to speciation. This simple argument also had tremendous philosophical and religious repercussions, hinted at in its conflicts. When Samuel Wilberforce asked Thomas Huxley on June 30, 1860 at the Linnaean Society from which ape lineage he originated, Huxley replied 'better an ape than a pastor'. But *On the Origin of Species* is not Darwin's only work, and it constitutes 'one long argument' over a total of three books.

Upon its publication, the last line of the *Origin of Species* suggested the future appearance of a work on discussing its implications for humanity—which Darwin finally publishes twelve years later in the *Descent of Man* (1871). The book treated the evolution of man as any other animal species. In it he added the role of sexual selection while deemphasizing that of natural selection—a view that would not be shared with Alfred Russell Wallace. Darwin's *Expression of the Emotion in Man and Animals* (1872) was originally going to be included in *Descent*, but its length implied its own independent volume. To understand Darwin's contribution, all three have to be read.

It is somewhat impressive to consider how early his foundational ideas were formed. In 1838, Darwin comes up with notion of natural selection after reading Thomas Malthus, and a year later he comes up with the key notion of human *moral sense*, as a reaction to James MacKintosh's book. Darwin had been profoundly affected by the ideas of his social milieu, and in 1827 met MacKintosh, a friends of Darwin's uncle and in-laws. MacKintosh left a deep mark on Darwin, who at the time knew very little social or political theory and absorbed all that he could from the elder professor. MacKintosh was married to Harriet Martineau, an independent woman who wrote and made her ideas known as an early feminist prototype. Journalist, editor, and thinker, Martineau publishes *How to Observe: Manners and Morals* in 1838, arguing that all cultures have distinct ethics according to the diverse circumstances of their own formation, while at the same time recognizing common elements to all.

These ideas were thoroughly elaborated by her husband in his *Dissertation on the Progress of Ethical Philosophy* (1836), which made a key distinction between *moral sense*, an ethical feeling, and the *moral criterion*, or the criteria by which to judge actions. While the moral sense was universal in humanity, the specific moral criteria varied between cultures and societies. MacKintosh had been influenced by William Paley and Jeremy Bentham, who argued that what was good for individual was good for the community.[12] The question which naturally arose in this context was where the moral sense came from; where did it originate? Both Martineau and MacKintosh took a religious stance, arguing that morality was imbued to humanity by God—a view first proposed in the eighteenth century. The moral sense formed a part of divine providence. Yet, why should gut feeling necessarily coincide with the general welfare? As it had been noted, only upon later reflection do we realize that heroic acts are good for the collective, but it was equally clear they usually were not the product of reflection at all at their moment of expression. There need not be any correlation whatsoever, and we might observe that many gut feelings may just as likely result in a criminal acts, as is typically the case when a husband's jealousy leads to the murder of his beloved wife.

This was a question ideally suited for Darwin, as there are a couple of common elements with his evolutionary theory. There is the notion of a great deal of variety (moral criterion) within a broad uniform pattern (moral sense), accounted for in this case by

---

[12] Bentham's utilitarianism defined the public good by that which was counted as the greatest good; the more individuals benefited by a particular policy, the more correct such a policy was for society.

theology. The problem was hence well suited within general scheme of natural selection, which would be able to account for an underlying uniformity amidst apparent superficial diversity, if applied correctly. Darwin also realized that he needed to provide a naturalistic explanation, for otherwise religion would ultimately creep back into his theory of evolution. Solving the problem of moral sense would help to reinforce the underpinnings of his biological revolution.

It was clear to Darwin that sociability provided a key to the evolutionary success of certain groups. Belonging in a group endowed individuals with distinct benefits, as protection from predators in the form of warnings of nearby threats or in a common defense. He provides a couple of specific cases. In one instance, Darwin describes the story of a baboon troop crossing a small valley. The leaders of the troop had already walked up the hill, while most of the group was still in the valley when attacked by African wild dogs. The male leaders with large canines flank down and scare away the dogs. As the troop resumed its march, however, one small trailing pup who had his lost parents during the attack was left behind. The dogs attacked again, and the abandoned orphan climbed onto a rock. As he was completely surrounded by wild dogs, his chances for survival were slim. A dominant male from the main pack proceeded to calmly walk down to the orphan, quietly passing through the wild dog pack, and encouraged the frightened orphan to join him. The alpha baboon successfully took him away from what was likely an eminent demise, protecting the orphan at a great potential cost to himself.

The lesson in this dramatic example was rather clear to Darwin. Groups where self-sacrificial behavior was predominant would be much more successful than those where it was not, specifically the case of solitary animals. Because social groups have a higher likelihood of surviving, such traits would eventually spread over evolutionary time. It goes without saying that the dynamic operated at the group level, and is referred to as altruistic behavior. Yet self sacrifice had an obvious and clear cost to the individual. For Darwin, if natural selection could act on physical traits, it could also act upon psychological and mental traits as well, specifically morality and culture, as in the case of altruistic behavior. Self sacrifice was adaptive because it conferred greater survival to those particular groups which demonstrated it, which in Darwinian evolutionary terms meant greater brood, and descendants. Darwin postulated that a similar phenomenon occurred in man (humanity).

Man's higher mental powers led to an advanced sense of sympathy towards others, which in turn led to a concern about the opinion of others (public opinion), and eventually, through long cultivation, to a sense of individual self control. For Darwin, the level of civilization generally dictated the range of this notion. Primitive savage tribes only showed sociability internally, for example. Truth telling was enforced locally (within the group) while lying and cheating to someone from another group was not only justified but actually favored—particularly in the case of interactions with rivals and enemies. Truth, under this schema, was not a universal human trait but socially circumscribed. Ethical notions were thus not absolute but relative to a person's reference group, again depending on the level of civilization of the group.[13] The moral sense of savage communities was restricted only to members of the same tribe, while advanced civilizations retained a more broadly abstract moral sense.[14]

Did animals then also have a moral sense as well, akin to that of savages? Darwin claimed that they did not, as they were subject to the whims of instincts without reflection—as in the conflict between maternal and migrating instincts. At times during the year the two coincided. Darwin noted that many birds abandoned their eggs and chicks without a second thought— which indicated the strength of the migrating instinct in birds. Caged birds would begin to beat chest against cage and bleed during these times of the year as well. Did the bird reflect upon

---

[13] There is a curious similarity between truth and warfare. In the primitive context, in-group warfare is rejected, while it is justified in out-group conflicts. The same dynamic occurs with truth telling, hence accounting for the strange anomalies observed worldwide (i.e. the lack of consistent behavior across cultures.).*

[14] It is not being claimed that the observation can be used to interpret Puerto Rican politics.

taking a thousand mile flight that their chicks were being left behind? Did this action constitute a moral tragedy for the bird? While we do not know, it is unlikely. The key difference between humanity and animals resided in the power of reflection.

In contrast to other animals, humans were always recollecting past and present experiences, and evaluating themselves on the opinion of others. The incapacity for animals to compare past and future actions meant that were not capable of moral reflection, according to Darwin. However, he also believed that any social animal with elevated intelligence would tend to develop a moral sense, regardless of what the species was; but it would likely be very different from our own.

There was a key conundrum in Darwin's theory, however. Does group survival justify morality at all? If the cost of group survival to an individual is too great, they will have no incentive towards this behavior.[15] Since evolution is not goal oriented, it is 'non-teleological', then there is no reason why self-sacrifice should have emerged in the first place in classical evolutionary theory. The individual who sacrifices himself for the group will likely not leave any children behind, and thus such a trait would not spread within a given population. Self-sacrificial behavior is ultimately nonadaptive for the individual, who is more likely to die prior to passing their genes onto future generations.

There is an inherent conflict between natural selection and self sacrifice that was not adequately reconciled by Darwin, and hence why the problem of altruism is such an important topic in Darwinian theory. It represented an 'anomaly' in Thomas Kuhn's terms. The moral sense of man was actually one of the topics Darwin thought a great deal about, yet one about which he actually published very few pages—akin perhaps to Newton, who had a vast corpus on writings on alchemy but seldom put such ideas to print.

Ethics was Darwin's alchemy.

---

[15] Such rationalizations have been used in the past as ideology for repression.

Is Human Nature Good or Evil?

What is the character of human nature; are we inherently good or evil? This apparently trivial question has enormous repercussions, from child rearing to public policy. According to the Judeo Christian tradition, the mind or 'soul' is an ethereal matter separate from the body. Humanity is inherently 'bad' due to original sin, and as a result, educational policy will have to be harsh and severe. Punishment will be part of the regular ethical code, as can be seen in female orphanages in Puerto Rico during the latter 19th century, which forced "bad girls" to sit on their knees for extended periods of time. In this traditional worldview, parental care is defined by a generalized aggression to help curb inherent evil drives in the child; hitting a child was regular part of its socialization. We might observe also that the soul existed independently from the body; "John's head" can be separated from "John" without altering his 'soul' in any way. As noted by Daniel Dennett, brain donation is the only case of organ donation where donor gains greater benefit than the recipient.

It goes without saying that religious explanations for human nature and behavior are no longer credited as valid. We do not accept the notion of soul as a valid scientific entity, nor does the idea that 'original sin' constitutes an acceptable account of human nature. However, these explanations are historically important, and still affect the beliefs and outlook of millions of individuals throughout the world, in spite of a decline of Catholicism during the last century.

Contemporary notions have been most influenced by the 'blank slate tradition', or John Locke's *"tabula rasa"*. This notion holds that human nature has no essence and is ultimately a byproduct of experience. We are evil or good only in so far as a result of the experiences we have had in life. The blank slate definition of human nature has been a very liberating notion that has given a great deal of power to social institutions and public policy, resting upon the presumption that these will ultimately shape the individuals in a community. According to this theory, we can literally build society from the ground up depending on the institutions that are designed, and it is easy to see why it was the dominant theory during much of twentieth century.

At the cultural level, it is best exemplified by the anthropologist Franz Boas who sought to disprove that cultural differences were due to racial causes. Boas was fighting racialist theories of the nineteenth century used to justify repressive policies; for him, differences in culture were due not to inherent human limitations but were simply the outcome of experience. His greatest influence was through his students, particularly Margaret Mead whose *Coming of Age in Samoa* (1928) portrayed an idyllic paradise of noble savages where uninhibited premarital sex abounded. One cannot overestimate the enormous impact of Mead's work, constituting apparently irrefutable 'proof' of Boas's ideas. Through it, Mead became a powerful figure in anthropology; a 'sage' who believed unequivocally in the blank slate model. She helped in the establishment of a 'feminist' view of the world, where anybody could become anything if only given proper care and attention.[16] Mead routinely participated in the United Nations, and her *Cultural Patterns & Technical Change* (1953) was a guide on the introduction of new technologies to old cultures, while aiming to prevent negatively affecting their mental health and social constructs. Culture, under her Boasian rubric, was defined as an intricate web; alter one part, and another would be severely affected. Culture was akin to a glass figurine, which could all to easily be broken, and hence required specialized and delicate care.[17]

Her figure is similar to that of Ashley Montagu's who also drew up proposals for UNESCO. While rejected initially, his

---

[16] As experience is ultimately definitive of the construct of character within its theoretical framework, there emerged an overzealous concern with regard to the negative exposures of the individual, which mirrored broader cultural dynamics.

[17] Incidentally, my father Gonzalo Fernós López debated her at UNESCO, criticizing how few dared to challenge her claims.

ideas were gradually incorporated into its creation. It goes without saying that Montagu had also been a student of Boas as well. Another such student was Elizabeth Marshall who studied the !Khung in the Kalahari Desert. Marshall also discovered a peaceful idyllic society, and as Mead, saw only what she wanted to see given that the actual rates of violence were akin to those of inner cities in the United States. Mead for her part had been deceived by her informants. Interviewed 50 years after the fact, they admitted having lied to her by painting a false picture of their culture's sexual rites and practices. Premarital sex was punishable by death in ritualized sacrifices. Mead's notion of an uninhibited sexual paradise was patently false, but would have enormous repercussions in social theory for more than half a century. The rates of violence in hunter gatherer societies reveal far different entities from those portrayed in noble savage depictions. Ninety percent are found to have engaged in some type of warfare, 64% routinely did so every two years. While aggression does not in and of itself discredit the blank slate, it does gravely undermine the myth of the noble savage.

Part of the problem with Boas's theories was that they were initially meant to account only for cultural differences between social groups on the basis of experience and context, rather than inherent biological limitations. For example, that many tribes do not have numbers past the number "5" did not mean that they were intellectually inferior to Western Civilization. In fact, when exposed to modern mathematical concepts, these are readily adopted. An Isaac Newton in African Sub-Saharan tribe would not have invented calculus or law of gravity not for lack of intelligence, but due to his social circumstances, as we have noted previously.[18] Given that mathematics is inherently graphical, the absence of a written language would put a damper on such efforts.[19] The specific problem with Boasian blank slate

---

[18] Rodrigo Fernós, *From Galieo to Boltzmann: A History of the Fragility and Resilience of Science* (Corpus Christi, TX: VirtualBookworm, 2016), passim
[19] Newton was greatly influenced by store-bought books given the thriving book trade in England; and his work can be seen principally as a reaction to Cartesian philosophy.

theory is that the notion became overgeneralized to account for all human behavior on the basis of experience alone. This meant that men were infinitely malleable; the focus on cultural variety which had stimulated the original theory was lost over time.

Given that the theory became the dominant principle in social sciences and social policy in the United States, it set the model for elsewhere and lent itself to utopian thinking. Social planners could reshape man and the world at will; that 'man' (humanity) could take any form led to quite unrealistic policies globally. These postulates were even adopted in China by Mao Zedong and his Cultural Revolution of the 1960s. Obviously human nature is not as simple as that, but such ideas underlay long-lasting and powerful social institutions globally. The prevalence of the Catholic Church through Latin America had an enormous impact on its social structure and organization, and also tended to view behavior in very binary and simplistic terms. Ironically, a system that was designed to institutionalize ethics, led to a culture of corruption, due in part to an ahistorical view of the world.[20]

The Blank Slate model has been predominant in the West throughout much of the twentieth century, serving as the underlining philosophy for numerous social institutions.

It is to be noted that a whole host of fields impinge on notions of 'human nature', aside from biology proper. At least 20 research domains covering 10 different disciplines have something to say about the topic.[21] However, there is a noxious puzzle in the analysis of the topic: what we believe of ourselves affects our behavior, as a mirror image of itself leads to infinity, and constitutes a subset of what we believe of world and nature.

Our beliefs have a tremendous impact on our behavior. If, for example, we believe that the end of world resides past the rock of

---

[20] The Spanish crown did not pay officials, who were thus force to 'tax' to make money, which became institutionalized as its current 'culture of corruption'. One cannot drive in Mexico without a policeman asking for a '*mordida*' (bribe). Puerto Rico's culture of corruption is present and much more subtle.

[21] These include, but are not limited to, animal behavior studies (primates), archeology, anthropology, behavioral genetics, neuroscience, cognitive science, child development, evolutionary psychology, sociology, political science, and behavioral economics.

Gibraltar in the Mediterranean, then we will limit our attempts to sail forward beyond these.[22] Similarly, if we believe ourselves to be inherently evil, we will likely behave in an 'evil manner' or we will be more prone to rationalizing evil acts—an all too common human trait. During the 1990's there was a sudden drop in crime in the United States, which was hard to account as criminal policies had not drastically changed. A more comprehensive study revealed that the trend was due to the US Supreme Court ruling *Roe v. Wade* (1973), making abortion a constitutional right for women. Unwanted children tend to be mistreated by their parents, who in turn internalize/externalize view, leading to a high likelihood of criminal activity in a society. The fewer the number of unwanted children there are, the lower the rate of criminal behavior will become.[23]

However, if we believe ourselves to be good, we will be more likely to behave morally, and this is actually a key principle in the US constitutional order. All leaders seek to retain a positive public persona. If moral stances are claimed in public by politicians, these will be implicitly pushed onto public moral behavior in order to maintain a positive public image, as clearly seen in the case of George Washington. In this sense, public positions of morality are self-reinforcing, and encourage political moral conduct in a virtuous self-reinforcing circle; if politicians fail or are exposed, they will consequently lose political power in a rational democracy.[24]

This is a notion of virtue different to that found in Shakespeare, where virtue equated being true to oneself; in the US Constitution, virtue means being true to others and in keeping one's word. Men acted virtuously by the sheer thought of how others would respond to a particular ignoble action; the US

---

[22] This is why Columbus's trips were so historically significant. Columbus was breaching a '*ne plus ultra*', going beyond the limits of the Greeks. Columbus in fact believed world to be smaller, thus set off on journey believing easier—a factual mistake which had good results, never realized discovered a new continent.

[23] Incidentally, the statistic also shows how important family relations are in formation of individual.

[24] Forrest McDonald, *Novus Ordo Seclorum* (1985).

founding fathers did not presume men were vicious, recognizing the noble passions which led to virtue, as those stirred by the family. Washington actually detested public posts, given that errors would ultimately affect his public reputation, of which he was so zealous in maintaining. He was so fearful of ruining his reputation, that after 4 years he left his post, thus accidentally establishing of one of the most common political cycles in the world.

Inversely, if others believe we are evil, their reaction and expectations will likely influence and promote this behavior—as the recent case of the Tennessee policeman who believed black drivers showed a much higher propensity to shoot, and tragically ended killing an innocent civilian on an ordinary day's work. Belief in aggression has a positive feedback dynamic. While initially a reputation for aggression will lead to a reduced frequency of personal physical attack, it can also increase the severity of any future attack. The important point, however, is that beliefs will frame how we perceive behavior. If all men commit unjust acts, then they will likely rationalize their behavior, thus tending to excuse away immoral conduct, pointing to a key problem in biological explanations.

A scientific rationalization for bad behavior thus acts as a triple whammy, and is known as the 'naturalistic fallacy' whereby prescriptive 'ought' claims are confused for descriptive 'is' claims, inevitably landing in the confused tautology of circular argumentation. Claiming the world to be as is, justifications are created for the existing status quo, akin to 'water is wet because it is water'. The circular argumentation is fairly obvious: we are evil because it is in our nature to be evil, and hence description turns into prescription for our becoming 'evil'. That being said, we should be most concerned with simply answering the question of its validity. Ultimately, ethical decisions and evaluations are influenced by this perception.

There is sadly a long history of scientific rationalizations for sociopolitical repression, which should always be kept in mind so as to not repeat the mistakes of the past. The most obvious example is in the justification of slavery. Certain racial groups are 'animals' and hence inhuman; any notion of civil rights would not presumably apply. Phrenology at one point showed number of lumps on head to be indicative of intelligence and character. As humorously noted by Steven Jay Gould, it was found that Georges Cuvier's hat was extremely large, which seemed to reinforce the predetermined conclusion prior to the study: larger brains created more intelligent creatures, thus proving Cuvier's brilliance.

There are other cases where science or natural philosophy became an instrument of colonial ideology. Pre-Columbian indigenous societies were rife with these philosophical-colonial dynamics. Aztec physics was used as a justification for the brutal occupation of nearby tribes, and ultimately in the creation of the Aztec empire. Because the notion of inertia did not exist, it was claimed that in order for world to move, human sacrifice was required. Inca biology was also used to justify the extension of prior chiefdoms. The bones of prior Inca rulers were carried and actually consulted, operating on the presumption that long ago deceased ancestors were still alive. This belief, however, allowed the establishment of an untouchable system of taxation rights for their heirs, which forced new rulers to occupy new territories so as to replenish the Inca state's source of income—and hence territorial expansion.

The European conquest of the Americas was also characterized by these dynamics, specifically the theory of degeneration or the notion that tropical climates degraded men and beasts, lowing their animus, activity, interest, reproduction and intelligence. The eugenics movement in the US (Cornelius Rhoads) similarly noted that too much sexual activity led to overpopulation, and hence required the forced sterilization and population control.[25] Science can provide both the justification and the instruments of repression.

As shown by Daniel Headrick, science can become the quintessential colonial enabler. The use of quinine, for example, allowed the conquest of Africa during the nineteenth century, as tropical diseases of the sub-Saharan region during most of

---

[25] While the Puerto Rican populational growth rate in the 1930s was high relative to US, this rate became the global norm after WWII.

colonial history prevented European incursions and created a natural barrier of entry. One of worst was malaria, whose cure with quinine allowed the partitioning of Africa between European powers as the cutting up of a cake, without the consent whatsoever of its inhabitants. Up to end of 19th century, Europeans had never seen a gorilla before. Malaria also played a big role in Caribbean history as well. Toussaint Louverture's rebellion during the late 18th cent in Haiti was successful due to it. Napoleon sends General Leclerc with thousands of hardy troops, only to be felled like flies due to yellow fever—a historical fact oddly disputed by the historian Gregory Cushman. Malarias and mosquitoes allowed Haiti to declare independence in 1804

Whatever our argument, we have to be aware of the tendency for groups to find convenient justifications for repression. This is particularly tragic when science become a part of these processes, providing the added social weight of the appearance of truth without actually being so.

Yet, what is human nature? It is a tricky question to answer.

History of the Blank Slate

Our ideas of human nature depend to a significant degree on whether we believe it to be innate or constructed, or what is usually referred to as the 'nature / nurture' debate. For much of modern Western intellectual history, the prevailing view has held to the latter, in what is also referred to as the blank slate model, which rose to prominence during the twentieth century. We will briefly discuss its history, and note that while many today still seek to prove it scientifically, the model is problematic.

It goes without saying that we have to be very critical of our causal presumptions, as these will certainly affect our judgment and conclusions, perhaps negatively. In *A Treatise on Human Nature* (1740), David Hume noted that we often tend to infer causation from sequential events in time. If something follows another, we tend to assume it as its cause: if A => B. I hit a ball, the ball flies through air, hence I am the cause of the ball's flight. This is also what Aristotle referred to as *efficient cause* or the proximate cause, immediate precursor to an event. For him, a creator (the Greek version of Christianity's 'God') was the proximate cause of motion in universe, as it was the first to turn the aether (external sphere), which in turn affected all the crystalline spheres below it, on which the rest of the celestial orbs moved. The universe was a huge mechanical clock, which was first set into motion by God, as noted by Henry Moore.[26]

For Aristotle, everything in the universe had a purpose and role or function, now defined by the term "teleological". It provided an awe-inspiring magical sense of the world, which has

---

[26] For Aristotle, there were four principal causes in the universe: 1) material cause or the wood of chair, 2) formal or the chair's structural form allowing a person to use it for sitting, 3) efficient cause previously seen, as the carpenter and 4) the final cause or the purpose the entity served, in this case, an object for sitting.

mostly been lost today, akin to she shift from the *sacred* to the *profane* described by Mircea Eliade. It goes without saying that it was an anthropomorphic view of the world: nature implicitly showed the same purposeful behavior as humanity.

Yet the truth of the matters is that correlation is not necessary causation. Just because B follows A, does not mean A caused B. The two might just have accidental circumstances, or a noncausal correlation. For example, if I sneeze prior to a car crash, this obviously does not mean that I caused the car crash.[27] Every time I sneeze, there will not be a car crash. The street light might have been out from a blackout due to Hurricane Maria.

This is obviously only a hypothetical example, but one which is much more common in social sciences than one might presume. We need to identify an event or object's mechanism in order to be able to identify the causal chain of relation. One view held is that we obtain the truth by mechanism, or what Giambattista Vico referred to as the *verum factum principle*. God knows the world because he constructed it; nature is ultimately known only by Him because of it. It follows that only when we build a model that reproduces a phenomena, can we then claim to understand the phenomena. In fact, this analysis/synthesis method is precisely what Isaac Newton used in his work: breaking things up apart (analysis) and then bringing them back together (synthesis). In his analysis of light with a prism he broke it up into its component colors, and reconstituted them with lenses. Similarly, humans can know society because they constructed it, according to Vico.[28]

The chain of causes might actually be much more complicated, possibly requiring a number of factors to all activate at the same time or in a distinct sequence. The greater the number of factors, the greater the types of interactions which increase in such an exponential manner that it becomes extremely hard to

---

[27] The arrogance of some drivers, along with the abundant non-functioning street lights, meant a very high number of accidents in the weeks that followed Hurricane Maria (September 19, 2017).

[28] These are but some of the debates referred to as epistemology in philosophy, or the concern with how we know what we know. What are valid basis for truth claims of the world?

fully identify the causal chain. For example, If $A1$, $A2$, $A3$ => B. In other words, for B to occur, all need to be activated at same time. However, B might also be caused by the activation of these factors in a particular order: If $A1$ > $A2/A3$ => B. A2 needs to be caused by A1, which in turn has to activate at the same time as A3, before B ever occurs.[29]

It goes without saying that solving the chain of causation gets very tricky when dealing with an exponential number of factors: If A^X => B, where x might equal 16,000 ($A^{16,000}$ => B) or a product where 1 is followed by 4,800 zeroes. "Tricky" however is an understatement, as it is extremely difficult to isolate parts and identify interactions with phenomenon of such order of complexity. It certainly cannot be "mentally" determined; one cannot just sit down and 'think on it'. The human brain can at most deal with interactions of 150 parts[30], and is not equipped to deal with larger datasets. Here the use of computers becomes a fundamental tool, in a discipline is referred to as bioinformatics or the application of computer science to biology.[31]

The Blank Slate model has a lot of problems qua scientific theory. It culminates in BF Skinner's behaviorism 1950s at Harvard University, claiming that all animal behavior was solely the outcome of reward and punishment. This conclusion was induced from a study of pigeons and rats which were trained to

---

[29] The greater the number of factors, the greater their mutual interactions, and the greater the difficulty of their decipherment.

[30] This has been the estimated size of tribal community, as judged by the Dunbar number.

[31] Incidentally, the emergence of modern computers was strongly influenced by biology, particularly the father of modern computing John von Neumann. His key computing infrastructure components as memory, input/output, and CPU were all abstraction of key human brain functions. ALL computers today share this basic schemata, described in his replication model of 1948, which inversely seems to have influenced James Watson and Francis Crick. For a self-grown computer to exist, von Neumann indicated that it needed to first make a working copy of itself within itself. The strong mutual influence between science and technology was first modeled by Edwin Layton in his 'mirror image twin model'. Prof. Layton, one of the founders of the history of technology in the United Sates, worked for many years at the University of Minnesota.

peck buttons or pull levers in order to obtain food. Any animal could be trained to do anything, claimed Skinner. However, when students tried to apply this to circus animals, they grossly failed. Raccoons did what raccoons do (washed the chips), as well as hogs who 'rooted' the chips, rather than inserting the chips. The experiment was a complete failure, thus showing that Skinner's work was not generalisable to other species. Animal minds are not blank slates, as if they were empty without any prior structure. This critique had actually been made by Gottfried Leibniz when John Locke first proposed his model, pointing out that the intellect is empty of sensations, except for intellect itself. Locke did not have a valid response.

Further evidence against the *tabula rasa* model of the mind can be tragically found in its therapeutic application, as in the case of David Reimer. Born as one in a pair of twins, David's original name was Bruce, and his brother's was Brian. The newly born twins were circumcised with a new chemical treatment of electrocauterization, which in David's case cut off half his penis. The doctors were unsure what to do, whether to keep the 'broken penis' or to completely remove it. The answer was determined by psychologist John Money, a strong behaviorist a-la-Skinner. Erroneously believing he could mold the human psyche into any form he desired, as if it were silly putty, he convinced the parents to remove all of Bruce's sexual organs. The testes were excised, and the genital area reconstructed. Bruce was turned into Brenda, both physically and psychologically in an extreme example of 'gender reconstruction'.

It was presumed by Money that, if treated like a girl, Bruce would become a girl in spite of having been born a male. His hair was cut and made up like that of a girl's; he was dressed like a girl, and so forth. Neither brother was ever informed of the traumatic event. Tragically, the experiment was a complete catastrophe. As a child, Bruce did not fit in with either gender. On the one hand he was too aggressive when playing with girls, and on the other he was rejected by boys for being a girl, as boys are wont to do. Bruce ultimately committed suicide at the age of 38. He eventually did discover the secrets of his past, and at one point lived as a man, but had a great deal of trouble coping with society, being maladjusted from his early childhood experiences. There are serious flaws with the *tabula rasa* model, which is still often presumed as valid within many fields in the social sciences—particularly feminist studies.

It was later discovered that testosterone strongly affects prenatal fetus, specifically its neurological development. It is for this reason that gender cannot be randomly reassigned at will. After conception the amount of testosterone in the mother's body also has a neurological effect. Women living in conditions of high duress, as those during circumstances of war, tend to have a higher incidence of gay children.

This is not to say that the blank slate view should be discarded completely. It has also been shown that experiences prior to puberty play an important role in shaping gender identity. The lack of rough tumble play with other boys, or the absence of a close contact with the father, while at the same time having an overbearing relationship with the mother, strongly contributes to the identification of the individual as a female, regardless of actual physical genitalia. By the time the individual reaches puberty, they identify themselves as having a female identify. The "sexual sense" remains the same, akin to a 'moral sense'; the individual will have sexual desires which, however, are fixated to nonreproductive ends. Male homosexuals are tragically 'bred' by the very fathers who reject them, eternally seeking for that emotional bond through a physical means.

Humans are distinctive from other animals in that child development is extremely long, hence the slow socialization of individuals will lead to slow and gradual mental and psychological formation. By contrast, the monkey brain is essentially set at birth, as in the case of chimpanzees. During this long human socialization, however, irreversible psychological properties arising from personal experiences and contingent on accidental circumstances, are eternally set. Some of these contingent factors include the parental personalities or the number of siblings.

Is there a contradiction between the two views? Not necessarily. Rather, as we noted previously, it suggests that the human psyche is the result of complex interaction between genes and the environment. Does this mean that behavior ultimately reducible to genes? Again the answer is extremely complicated. There is no simple no simple A=>B, as had been presumed by Nicholas Brokaw. In 2001 the *Human Genome Project* was completed in record time by Greg Venter. One of its surprising findings was that only 34,000 genes were active, a much smaller number than was previously believed (150,000 genes). Venter oddly concluded that the genetic determinant interpretation was incorrect, thereby reinforcing the blank slate model and the behaviorism it implied. This bizarre and odd claim might have been due to the political pressure created by what that the *Human Genome Project* potentially implied: a deterministic model of human nature. However, it goes without saying that Venter's conclusions were incorrect.

One might ask, as Seven Pinker does, what is the number of genes that a deterministically genetic behavioral theory would require? The state of genetics is still in its infancy, and there is a question as to whether Venter's 34,000 figure is correct. These were identified only by the known proteins made, which by definition is incomplete.[32] Regardless of the case, there can be no doubt that the complexity of an organism not determined by its genetic complexity. The roundworm has 20,000 genes, but only 959 cells, while humans have hundreds of trillions of cells, but only 34,000 genes. In fact, a corn plant has more genes than a human being, but it would be difficult to argue for a corn plant's greater complexity. Can corn write a breathtaking opera, paint a sublime landscape, or compose a rich novel?[33]

Single genes also do not necessarily correlate with any one particular feature; 30,000 genes are not correlated with 30,000 traits, as genes also interact with one another. In many cases, for one gene to act, it must be preceded by another, thus exponentially increasing the variability of outcome. The more accurate number of interactions is not 34,000, but rather in the range of $X^{16,000}$, or 1 followed by 48,000 zeroes. Chromosomes are a long lists of genes, where different areas mark for different

---

[32] It is a number that has to be accepted only skeptically.
[33] Movie *I, Robot* (2004).

proteins, and in turn for different genes in the same strand. Given this complexity, it is clear that genetics is still in its infancy, and will not be wholly identified and mapped out for some time to come—and much less have its relation to behavior easily plotted out.[34]

The best way to understand the nature/nurture debate is by looking at the implicit debate between Hobbes and Rousseau in its own historical context. Is man good or evil in a state of nature? Why? The answers provided by both philosophers had an enormous impact on Western intellectual history and its institutions.

Thomas Hobbes believed that men were evil in a state of nature, which implied that each were at constant conflict with each other, *bellum omnium contra omnes*. As the cliché goes, life was brutish and short. The bases of civilization as commerce and knowledge could not flourish under these circumstances, and hence required a powerful state to insure their survival, civilization, and prosperity. Hobbes referred to this state as *Leviathan* (1651).[35] Hobbes did not define the nature of the governmental leviathan in that the state could be a monarch or a commonwealth. However this state needed to have absolute power over all others to prevent the decay into barbarism; nothing could be above it, and thereby provided the modern definition of sovereignty. For Hobbes, only by force (or the fear of force) could order be established and maintained in a society of men. Men by definition inherently sought power and persisted in their ambition only until death.

By contrast, Jean Jacques Rousseau was an anti-philosophe who believed that man was inherently good in a state of nature, alluding to the noble savage. In contrast to Hobbes, Rousseau believed that society had a noxious effect on the human psyche. Men were turned evil by the societies in which they lived. He disregarded the legitimacy of national government as any ruler

---

[34] While we might be cynical at this fact, it is actually positive, given that scientific knowledge tends to be abused by nonscientific actors.

[35] While we might presume that he alluded to the horrific sea creature akin to that seen in the movie *Pirates of the Caribbean*, in fact it was only a whale.

will only be interested in their own personal power rather than the common good or public interest. This in turn implied the need for a *Social Contract* (1762) to tightly bind men together. Here Rousseau is not alluding to a regular contract that can be casually entered to, and encapsulated the 'Blank state' position in that men are ultimately shaped by circumstances. For him, not only was society inherently evil, but he also believed that science and art led to the corruption of morals.[36]

Hobbes and Rousseau are obviously at diametrical odds with each other. Man was good in one, evil in the other; society necessary for one, injurious in the other. Both were extremely influential as political theorists. The French Revolution was inspired to a degree by Rousseau, but he was not the only factor in its chaotic history (1789-99). It position with regard to science was erratic, as in many other things. Lavoisier was killed with doctor Joseph Ignace Guillotine's invention. Robespierre who briefly led the Revolution, read Rousseau's works, which in turn helped him to establish key points in the *Declaration of the Rights of Man*.

During the nineteenth century, the influence of Hobbes had not only waned but tended to be dismissed. Hobbes had taken a 'realist' point of view. While is normal in our time, it shocked what was still a religious era which saw man as a creature of God, and in him the spirit of God as well.

While not the most popular of philosophers, Hobbes however did create one of the most influential political theories, specifically the notion of national sovereignty: a principle held by every nation today, in theory if not in practice. It goes without saying that Hobbes's political theory is the more widely accepted one today, and it accurately captures the current state of world affairs since it was written. With regard to the needs of the state, might makes right. The dominant political view is still that of political realism, whereby no overarching body is currently above the nation state, leading to a chaotic mix of sovereigns in

---

[36] Ironically the *Social Contract* was an essay written for a contest held by the Parisian Academy of Sciences at behest of Diderot, who obviously disagreed with its conclusions.

international affairs.[37] The first League of Nations was powerless when first formed, and the United Nations still has little real autonomy; its leadership by consensus is subject to the influence of global superpowers.

A matrix of individual versus society on one axis and barbarism versus civilization on the other can be created, which summarizes the above points in a simple table:

	Individual	Society
Barbarism (immoral)	Hobbes	Rousseau
Civilization (moral)	Rousseau	Hobbes

To better understand these concepts we also have to look at their historical context, specifically the Voyages of Exploration (Era of Columbus) which so profoundly influenced both men. The Voyages of Exploration and of the discovery of the Americas had an enormous impact on the Western worldview. They exposed the 'European mind' to a whole host of new social and biological forms: new types of human societies and new types of animals and plants that did not fit into existing models. It turned out that the Greeks were not as perfect as had been previously believed. Most animals of the New World had not been described by Aristotle. More importantly, the enormous diversity of new animals and plants drastically expanded the diversity of known facts and ultimately led to their reconceptualization.[38] A wider sample selection always leads to more accurate generalizations in any field of study.[39]

---

[37] This helps account for the difficulty of obtaining effective substantive action on international issues, as with climate change; there is no international body above the nation state.

[38] It is only after Darwin traveled to South America that he formed his revolutionary ideas.

[39] These discoveries, however, had to be gradually assimilated—descriptions gathered, classification models established, and so forth—before their intellectual impact occurred. At first there was a great deal of awe and wonder, which gradually disappeared as more rigorous comparison studies were made.

24

Travel description literature was ravishingly consumed, and influenced many major thinkers. Michele de Montaigne, a wealthy diplomat who established a new genre of the biographical, depicted the world from his own personal point of view. His piece on *Cannibals* alluded to the notion of the noble savage, and played a role in both Hobbes and Rousseau. Incidentally, there is an interesting academic debate in Puerto Rico on the issue of whether *Caribs* and *Tainos* can be distinguished from one another, in which violence plays a central role. The anthropologist Jalil Sued Badillo in *Los Caribes* (1978) argues that no distinction exists between *taino* and *caribe*; for him all were violent, and the hence distinction is a false one, implicitly adopting the Hobbesian point of view of man existing violently in a state of nature. On the other hand in his *Sociedad de los tainos*, the historian Francisco Moscoso take a Rousseauian perspective, suggesting that the indigenous were inherently peaceful, and providing an over-idealized portrait of *taino* life.

The descriptions of Indians in Puerto Rico found in the *Crónicas* do reveal a great deal of violence and conflict. The *caribes* tended to come from the eastern Antilles. The eastern end of the island, where the municipality of Fajardo now exists, was routinely invaded. Other natives tended to flee from the caribs, and clearly cherished the brief protection Columbus afforded; islands as Culebra were not safe at all. The *Crónicas de Michoacan* (1493) are in accord much with what Hobbes had written.

For Rousseau man in a state of nature was noble because he was an autonomous independent agent. He might be poor, but he was his own producer of goods, which in turn made his social exchanges as those between equals. In this normal state of affairs, the expressions and demeanor of natural man were authentic. While in a state of society, men were interdependent. As all cannot produce all goods, social divisions inevitably emerged; which implied inequality, and in turn made nearly all of his social interactions unauthentic. All men in society wear masks, Rousseau insightfully noted. They pretend to be what they are not, and favor superiors whom they despise, for example. In the long run the theatrical character of interactions in a hierarchical society has a corrosive effect on an individual's character, whom like the statue of Glaucus becomes disfigured and hardly recognizable over the span of time.[40]

In *Emile* (1762) Rousseau thus presents education as the complete opposite to Catholic instruction. There were no punishment, but only rewards. The child is to be given a free reign to explore the natural world, which would develop what today referred to as 'agency' or the feeling of control over the natural world. Inversely, they were not to be given things to read for a long time.[41] There are obvious problems with this system, as with much else in Rousseau. His educational system overturned the natural social hierarchy, where a child might presume to know more than the teacher.[42]

By contrast, what was shocking about Hobbesian political theory is that no inherent immoral qualities were associated with evil. "Evil" for Hobbes was only the outcome of forces of social repulsion; there was no inherent immorality in evil, strictly speaking. We might account for this on the basis that he is ultimately a materialist, who believed that all could be accounted

---

[40] In this context, it is curious to contrast the traditional rural Puerto Rican *jibaro* to that of the modern average Puerto Rican. The historical *jibaro*, who is an extinct species today, was an autonomous fellow. During much of the colonial period most urban areas were not inhabited, and thereby, as Rousseau's natural man, produced a large number of his goods. His home was the humble *bohio*, or a shack on sticks; his way of life was based on subsistence living rather than the cash crop economy in the majority of cases. We cannot doubt his authenticity, even if uneducated. As witness by Bailey K. Ashford, his critique of the metropolitan '*letrados*' (learned ones) was not far off the mark; they often sought to take advantage of others, rather than enter into relations of equals that were mutually beneficial. By contrast, while the average modern Puerto Rican obtain the benefits of the city, the majority are poor (60%), live on food stamps, and are highly dependent on the welfare state—which in turn lead to question as to his authenticity and integrity (Rousseau).

[41] The Montessori school system adopted this philosophy. It is child centered, recognizing the natural curiosity of children. There is no curriculum per se, but is allowed to explore and discovery world, discovery.

[42] The biblical notion of *spare the rod, spoil the child* might have some validity to it.

26

for on the basis of atoms (corpuscles), and hence was also charged with atheism as well—then a serious criticism. Naturally, such views did not win Hobbes many friends. His turn to scientism came late in life.

In 1630, at the age of 40, Hobbes learned Euclidean geometry, which was something akin to a revelation. He saw that unquestionable truths could be established, and then sought to apply its method to the study of society. At the time England had been in Civil War, so Hobbes was forced to flee to Paris, where he met atomists as Pierre Gassendi and helplessly hears of King Charles's beheading. As a response, Hobbes sought to establish a political philosophy that would be in accord with the nature of man. His original aim was to proceed from the general to the specific, from the nature to man to the ideal political system, but was forced to write in reverse due to his circumstances, first writing on politics to then write on man and nature.

Hobbes noted that atoms in the world naturally cohere themselves into bodies when under the right circumstances, and believed that if he could similarly establish the right political rules, he would help establish peaceful and civil societies. In his modeling from natural philosophy (today 'science'), he equated man with atoms, and human will (motivation) with *conatus* (motion). Both conatus and human will were the sources of movement in the universe. Just as atoms were constantly under forces of attraction and repulsion, humans similarly were constantly under similar forces as well, specifically in his case pain and pleasure. All men tend to seek pleasure and avoid pain.

Again, for Hobbes, men in 'state of nature' constantly needed to obtain power because they never knew how long they would be able to keep it. Social power was needed to obtain those things which gave pleasure, and without which they would not otherwise obtain. As there was no overarching political entity or system, each individual tended to enter a state of warfare so as to obtain and retain the power necessary for the obtainment of pleasure. In this state of nature, there was no notion of property rights for example, and thus were constantly subject losing what today we regard as 'personal property'. By contrast, our modern notion of theft implies a scheme of values that was unknown to primitives, and has been widely commented in the European travel literature. Darwin noted, for example, that theft was common among the indigenous Fuegians of Argentina. Hobbes argued that such a chaotic state, or the absence of peace, prevented the development of the arts and crafts, or what today we would call scientific development. It is precisely for this reason that a leviathan was needed; for any hope of civilization his existence was necessary to impose order so as to provide the conditions in which civilization flourishes.

It is important to point out that in both Hobbes and Rousseau, the political state was the direct outcome of the nature of man. Just as atoms defined the nature of objects in the natural world, the nature of man had direct repercussions with regard to the establishment of society and the character of its people. It goes without saying that Rousseau disagreed with the majority of his colleagues, the *philosophes* as Voltaire, with whom he had a big falling out. Voltaire was an anglophile who brought Newton to France, and who ultimately perhaps placed too much emphasis on the role of reason in the understanding of human nature. To a degree, however, Rousseau was taking issue directly with the claims of Hobbes.

Ironically perhaps, Hobbes as Rousseau is ultimately also a follower of the blank slate. In both, the 'state of nature' did not mean 'human nature' proper but rather a pre-social condition or one prior to society and/or civilization. Note as well that for Hobbes there was an implicit commonality of man which allowed comparison between European and American to take place. The American Indian lacked civilization only because of the absence of philosophy, and he did believe there was no inherent faculty of mind that prevented them from its acquisition. Contrary to what one might suppose, Hobbes's stance is a not a racialist one. Only by establishing proper society, Hobbes believed, would men stop being evil, as they would have no inherent compulsion to do so.

In spite of all their differences, it is to be noted that for both, men's ethical behavior (or absence of) was the direct outcome of the institutional setting in which individuals inhabited. The key disagreement between Hobbes and Rousseau thus resided in their definition of the institutional setting required to meet this goal. For both, man was ultimately defined by his social context. A very different approach would be taken by the 'inherent qualities' theory, whereby the source of ethics did not originate in society, as the outcome of institutional arrangements, but rather emerged from within human nature itself.

Herbert Spencer and social Darwinism

In our previous chapter, we looked at the blank slate model: man was the result of his experiences. As a result, different mechanisms were postulated on how to establish moral societies. On the one hand, for Hobbes, only a powerful state (*Leviathan*) would lead to peace, which in turn was necessary for commerce and science. On the other hand, for Rousseau a 'social contract' was the key to a moral order, as it would limit the harms done by society to the individual. In both, however, the design of social institutions were essential in determining the character of men, and in turn the character of those societies. Both fundamentally shared the aim of establishing 'ethical' societies, or what we refer to as 'civilization' (civilized societies) which were relatively peaceful and truth-seeking as opposed to 'barbarism' (barbaric societies) marked by violence, fear, and deception.

In this chapter, we will look at the other side of the coin. Prior to the rise of sociobiology, there had been two substantive attempts to define an ethics from biology, principally that of social Darwinism and eugenics. While prior theories sought to establish a moral order via the modification of public institutions, these attempts sought to base their policies on a science-based understanding of human nature.

The ideas of the two movements were the natural outcome of their own presumptions. As noted by Bertrand Russell, we hold presumptions around us as flies; there are so many of them that often we do not perceive them in spite of their enormous impact, affecting our behavior and beliefs. They will inevitably have a profound impact on the ideas we hold about the world and ourselves. This is a natural cognitive process affecting all persons, regardless of culture, history or nationality.

Presumptions, by definition, are a set of beliefs implicitly held, and hence why the Socratic method was so effective as a type of 'scientific' method. Its elucidation of belief constitutes an exploration of the contradictions and consequences of a set of mutually ideas held.[43] Both social Darwinists and eugenicists believed that human nature could be used to establish a code of ethics. What could their preceding presumptions have been? To answer this, we must turn to the history of Western philosophy.

The eighteenth century was marked by a religious crises of sorts. Europe had seen the rise of new ways of understanding the world in what we now refer to as the Scientific Revolution. Their applicability to society, in turn, was explored in society during the Enlightenment in the works of Voltaire, Hume, Jefferson, and others. The *philosophes* of France were particularly active in the process, but had not been the only ones involved. In Germany, Immanuel Kant coined the term 'enlightenment'. There were consequent challenges to Christianity as a result, as many biblical truth claims were factually incorrect, if taken literally. For example, only one planet (Venus) is ever mentioned in the Bible. The *philosophes* were particularly critical of the powerful Catholic Church, and despised the Jesuits. A byproduct of the CounterReformation, the Jesuits were criticized for seeking to monopolize truth.

The challenges to the Catholic Church were not only with regard to its truth claims, but also to its moral ones as well. Christian fathers as St. Tomas Aquinas had built a house of sand upon the foundations of Aristotelian philosophy.[44] Aristotle's own father Nichomacus had been physician, and appears to have had a profound influence on his son. Aristotle names his most work on ethics after him (*Nichomachean Ethics*). To be ethical

---

[43] It is to be noted that it is an honest exchange of ideas, markedly contrast to the Sophist method where beliefs held prior to exchange. This is why the sophists were so detested by the likes of Socrates and Plato, in that they were inherently deceptive; their aim was to seek to get individual to change opinion as in court of law or politics. Unfortunately both approaches are often confused, as they use similar tactics and might be hard to distinguish to the untrained eye.

[44] Scholasticism, however, is not to be confused with Aristotle per se.

for Aristotle was to take the mean and avoid extremes in behavior. For example, what does it mean to be valorous or *'valiente'*? He did not take it to mean foolhardiness, or the going into problems without any forethought. This was short-sighed and could get one easily killed. On the other hand, it also did not mean no action at all, or what we would refer to today as cowardly behavior. Valour meant the realistic assessment of danger and proceeding to critically undertake a required action, as the killing a lion menacing a group. Cultural notions of manliness vary to a degree, and have been studied by various fields including mythology and anthropology.[45]

---

[45] In his *Hero with a Thousand Faces* (1973), Joseph Campbell notes that in spite of enormous variations, there is a common hero pattern of a going forth and a return. This action can be either purposeful or accidental. A soldier thrust into war may disagree with its purpose and ends, but still undertake a heroic ordeal. The hero journey takes many forms, as that of Prometheus exposing himself to danger upon getting fire or in the search for father, eloquently depicted in the movie Star Wars. In his *The Power of Myth*, Bill Moyers interviewed Campbell, who noted that all have to undergo ritual or initiation rite of some sort or other, whereby there is a shift from mentality of child (egoistically centered) to the mentality of adult which was (in theory) outwardly centered.

In his *Manhood in the Making* (1990), David Gilmore analyzed the many cultural variants of manliness. The rite of initiation for the Masai was to kill a lion while in Jewish communities it was the act of memorizing large sections of the Torah. It goes without saying that the criteria of manhood is determined by the requisites of group survival in the particular context of that group. Masai cattle herders have to regularly fend off lions; cattle provide physical nourishment in the form of milk and meat for the group, and hence essential to its survival. Similarly, since what it means to be 'Jewish' is inherently cultural by definition, their initiation act also serve to protect the group from outside threats, in this case cultural ones. Becoming a man often meant giving up the self in order to protect the group from outside threats, and which (by definition) goes against the instinct of self preservation. It is the natural tendency of all humans to run away when they see lion, for example; after the initiation rite, the individual now runs toward the danger for the ultimate benefit of the tribe. We may briefly observe that the definition of manliness is deeply associated with altruism; it is the *sine qua non* ideal definition of altruism.

The ritual for women, however, is very different. Femininity is biological and 'inherent', and thus the rituals are of a different nature. Is the 'female ritual' to protect group from inside threats? This is unclear. However,

It goes without saying that the Aristotelian worldview crumbled during the Scientific Revolution and the period of the Enlightenment which followed. During the early modern period, nature was now to be taken at face value. Scholasticism was epistemologically dead, and we see the beginnings of a broad pattern of secularization in Europe. The notion that religion could provide truths of the natural world was increasingly questioned, and became an attitude which also spilled into moral philosophy, with regard to the validity of its moral claims.

There were three general reactions to this cultural change, which ranged from denying its very existence to the attempts at modeling an ethics and a morality from the 'ground' of human nature.

The first type of reaction was to completely ignore the profound intellectual tectonic shift initiated by the Scientific Revolution, and can be seen repeated outside of Europe at a much later date; we might take Puerto Rico as an example. In the local literature of the colonial period, the Scientific Revolution is nowhere to be found. The discussion of physics often surprisingly refers only to Aristotelian physics, and it is somewhat shocking to see the enormous civilizational gap with the rest of world. During the nineteenth century, so abundantly dominated by Darwin's biological revolution, there was even less discussion of leading scientific ideas; and his shadowy presence shines by its overwhelming absence, or as noted in Puerto Rican slang, "*brilla por su ausencia*". This can be accounted for by Puerto Rico's colonial system.

---

it is interesting to point out that women are typically held as standard moral bearers across many communities, likely arising from their maternal roles providing unconditional love to their children. They in turn provide the definition of what a natural parent is. This might be contrasted to the historical depictions of 'unnatural parents' (stepparents), and their tendency of animosity towards the children. This animosity has been observed through the ages, in both literature and history. Cinderella is hated by her stepmother, and in the history of science, the rituals of alchemy could not occur in stepmother periods, which would otherwise lead to detrimental outcomes in the procedure.

As Alejandro Tapia y Rivera sharply pointed out in his autobiography, the policy of *"baile, botella y baraja"*[46] was common throughout the history of colonial administrations. Ignorant citizens are easily controlled, which led Tapia to gravely criticize the lack of educational institutions for four centuries, particularly his own 19th century. As there were no universities, higher learning did not occur. Most schools were controlled by religious authorities, as the *Seminario Conciliar* led by the Jesuits. Any Puerto Rican wanting to get a university degree had to go to Europe, Barcelona being a common site of study. Its high costs naturally meant that higher education was thus not 'universal', and only a few from the upper classes went to college—a lesson that has been forgotten today.[47] Tapia himself had been personally lucky in that his Spanish father had been military man and thus had enough wealth to attend, if just barely. However, Tapia ran into routine problems with the authorities in the island, whom he sharply criticized in his plays. The cases of other leading Puerto Rican intellectuals from the nineteenth century also illustrate the repressive nature of Spanish colonial policies.[48]

A second reaction was that of recognizing the conflict, but taking a circumscribed and limited response. The United States during the nineteenth century is an example. Thinkers in the US also initially ignored Darwin and the social implication of his *Origin of Species* (1859). However, this had not been a conscious

---

[46] Literally translated as "dance (festivities), bottle (alcoholic consumption), and cards (gambling)".

[47] Sadly, today the University of Puerto Rico is attacked by the very Board of Directors meant to protect its interest—blatantly disregarding the historical significance of its existence. Many government leaders do not genuinely appreciate what life would be like without it, and tragically seek to undermine it.

[48] Both Roman Baldorioty de Castro and Jose Julian Acosta received financial assistance at the behest of Padre Rufo to study science in Spain. Baldorioty was focused on botany and biology, and along with Tapia, Acosta, Betances and other Puerto Rican colleagues prepare the *Biblioteca Histórica de Puerto Rico* (1854). This was a compilation of documentation on Puerto Rico in Spanish archives, which shows the level of ignorance in which Puerto Rico was kept by colonial authorities.

decision, but simply the outcome of the difficult circumstances created by the contemporaneous American Civil War (1861-1865). Once the war was concluded, Darwin was avidly read and discussed, and the ethical-moral problem was immediately recognized: the epistemological credibility and moral legitimacy of religious institutions had been grossly undermined. This in turn meant an emergent moral and social crises for Christianity in the United States. There was inversely the belief that a science-based ethics would eventually emerge to meet these new moral challenges. Science was seen as key to any and all social problems. The main concern in this new context regarded the transitional period between the fall of a religious ethic and the rise of a scientific one—one that would be characterized by social chaos and disruption until a new 'world order' was established. Some pointed to antecedent historical examples, as Rome prior to Christianity, Hellenism after the fall of Greece, or Italy after the Renaissance.

A common response, however, was to adopt easy solutions to this dilemma, as was the case with Benjamin Kidd. A British clerk, Kidd produced a philosophical solution in that, from a strict academic standpoint, was unsatisfactory. However, his work was extremely popular as it hit many of the era's principal concerns. He wrote for those who feared the labor movement as well as the robber barons. While many admired calls for national duty and devotion, they were also horrified by the trench warfare and suffering of the Great War (WWI).

In his *Science of Power* (1918), Kidd noted the contradictions of *progress via competition*. Why should an individual sacrifice himself for the greater good if they were not guaranteed its benefits? He recognized that there were key conflicting interests between the individual and the society to which he belonged, and hence postulated that religion could fulfill such a function within capitalism. In other words, Kidd created a 'suprarational' justification for self-sacrifice; it was a religious impulse based on sympathy and modeled upon the ideal role model provided by women. For Kidd, women innately subjugated their present needs for the future benefit of the family, revealing a substantially different feminine culture from our own.

Kidd's response is a conservative one: retain the status quo with a new 'scientific justification' embodied in the key traits of Darwinian competition. It goes without saying that he also opposed socialism as it would eliminate competition, which in turn lead to the degeneration of society.

The third type of reaction is perhaps the most interesting. In it we see the attempt to create an ethics from human nature. It was not a passive approach, but actively sought out answers, and first emerged during the preceding century in Britain and can be seen as a precursor to 'moral science' in the literary movement of the time. Mary Wollstonecraft, author of the *Vindication of Rights of Woman* (1792) and mother of Mary Shelley (herself author of *Frankenstein or Modern Prometheus* of 1818), explored a 'science ethics'. Wollstonecraft attacked writers who denied the goodness of man, as the philosophers who were denying humanity to man himself.

The notion that a morality could emerge out of human nature itself could be seen in a variety of sources. The *First Encyclopedia Britannica* (1768-1771) defined moral philosophy as that which "traces from man's nature that which terminates in his happiness". A key notion of period drawn from the Scientific Revolution was the idea that a study of men would reveal God's purpose; as God had imbued man with goodness, man himself was a reflection of God. This idea is similar to Newton's study of the universe and discovery of natural law thereof: the discovery of the gravitation law that explained the coherence and motion of bodies, first set by God, was the discovery of God himself. As in Newtonian physics, it was believed that the study of human nature would also reveal the values intrinsic to humanity, and hence knowledge of what is right to do—what is essentially a scientific foundation of ethics. As in Newton, the study of man would reveal the laws of man, and hence the path to a harmonious social order.

For Francis Hutcherson, moral principles were deeply embedded into human nature; all men were endowed with an inner sense which allowed a person to perceive virtuous qualities of a thing or an action through the affections, akin to Rousseau's notion of feeling. Lord Shaftsbury's ideas about beauty are similar. We can identify beautiful things because there is an inner harmony in our minds, akin to harmony produced by the strings of a plucked instrument. When tuned to the right pitch, they resonate with each other at the right frequency. A similar phenomenon occurred in the human brain upon the perception of art. When the brain's internal harmony coincided with an external one in the painting, we obtained our sense of beauty—a notion which would be influential for Herbert Spencer.[49]

The first formal efforts towards a scientific ethics were undertaken by Leslie Stephen, whose good friend William Clifford had died at a young age after embarking on the same project. Stephen saw his efforts as a continuation of his friend's work, and produced the fullest expression of 'Darwinian morality' per se. Morality, as with any other evolutionary trait, is conditional to the surrounding circumstances of its holder. As a result, some moralities are better suited to certain conditions, while others might even be maladaptive given their contextual incongruence. In other words, for Stephen, there could be no universal moral codes, just as "hands" turned to wings for bats, and to fins for dolphin. As the physical bodies of individuals were always relative to their respective niches, so were their respective moralities and ethics.

There can be no doubt that the most influential social Darwinist, however, was Herbert Spencer. While Darwin sought to prove that morals had emerged from evolution, Spencer sought to determine precisely what its ethical tenets were, producing a comprehensive universal philosophy that sought to account for the entirety of existence. He had noticed that an increase in warfare was due to a decline in religion. Consequently, as eighteenth century moral philosophers, he recognized need to establish a 'science based ethics' for its substitute, which would become a social philosophy based on competition. Darwin was actually hurt by claims that his works led to social injustices.[50]

---

[49] Spencer is not to be confused with Francis Galton, whom coined the word 'eugenics'.

[50] In letter to his friend Charles Lyell, Darwin expresses his disappointment with the notion.

There was a deep ethical and moral strand underlying his character and work.

Spencer saw himself also as ethical a man as Darwin. Contrary to popular belief, he believed that men were inherently good, captured by the notion of 'beneficence'. As a consequence, men required freedom to express this beneficence, as long as they did no injury onto others. Spencer was equally opposed to war and imperialism, which he saw as a violation of the moral integrity of individuals. In contrast to Darwin, however, Spencer's vision was inherently teleological, believing that history was destined towards a particular goal. While his contemporary society was militaristic, Spencer believed that it would eventually result in an industrial utopia, characterized by peace and good will among all. This view is aptly captured in Aldous Huxley's *Brave New World* (1932). War was just a transitionary period to be ignored.

Spencer was an influential journalist. Editor of *The Economist*, he was also a member of the XGroup—a small but highly influential association in which he was an odd specimen. Unlike all the others, he was the only one that did not belong to the Royal Society, one of the leading scientific academies of the period. Spencer was also close friends with T.H. Huxley, Darwin's 'bulldog'; both were neighbors and would routinely go on walks to talk about issues. To top it all, Spencer was also a close acquaintance of Darwin's. Both had read Robert Malthus and had been influenced by it: while resources only increased arithmetically, animals grew at exponential rate , hence inevitably producing a merciless competition.[51]

Spencer's work was extremely ambitious, seeking to prove that evolution was the reigning paradigm of the entire universe, and not just that of life on Earth as in Darwin's case. He dedicated his entire life to this ambitious project. As Hobbes, Spencer initially planned to work from large topics to smaller ones, from the universe towards man and politics (ethics). However, when he suffered a sudden health problem, and became

---

[51] Malthus had also played a key role in the formation of the notion of natural selection, which so dominated the work of Alfred Russell Wallace.

worried that would die before completing his summa, he decided to reverse the order, as with Hobbes. He published *Data on Ethics* in 1879, a preliminary study describing his key ideas. Ironically, it turns out that Spencer need not have worried too much, living to the long age of 81 years. He was able to publish all of his works, which culminated in his *Principles of Ethics* (1893). He also wrote *Principles of Sociology* (1876), which is similar to the work of Eugenio Maria de Hostos in that both tried to derive laws of society from history, and in turn used these for the creation of the best political system, another example of the ambitious intellectual projects of nineteenth century thinkers.

Spencer's views were oddly based on a Lamarckian view of the world, which Darwin accepted at some point in his intellectual journey: the notion that acquired traits are passed down between generations. This Lamarckism endowed Spencer's work with a teleological character. Each generation saw a gradual improvement which, as in biology, was cumulatively added over time, resulting in its 'logical' utopian conclusion. Progress in his schema was an inevitable historical process, akin to the unidirectional function of the unidirectional function of the human vein.

Another important trait of Spencer's work was its 'long term view'. Spencer did not focus on immediate events, but rather sought to operate on a 'universal' time scale. As a result, immediate events for him where not necessarily refutations of his ideas, similar to the distinction of waves and tides. Wars were like small waves that did not necessarily mark the long term trends of history—a notion which influenced powerful political figures as Elihu Root.

Lamarckism, as one might suspect, became social Darwinism's Achilles' heel; when Lamarckian notions were disproven, the entire theory collapsed on its head. This occurred in 1890s with the development of biology, specifically August Weismann's germ plasm theory.[52] Note that Spencer wrote prior

---

[52] Germ plasm is not affected by the environment, hence suggesting that traits transmitted by both parents.

to the emergence of genetics; its impact was larger than even Weismann could have imagined.[53]

It is perhaps here important to observe that the philosophical impact of biological ideas. As previously mentioned, how we perceive the world affects the interpretation of ourselves, which in turn has social consequences—as was the case with social Darwinism.

Social Darwinism was principally a North American phenomenon, partly due to Darwin's greater acceptance. Asa Gray wrote the first favorable review of *Origin of Species* and became a Darwin 'guru'. Although the influential Louis Agassiz did not like the theory, his students widely read Darwin's work; his own son was converted to Darwinism against his father's wishes. Religious colleges in the US obviously opposed Darwinism. The reigning notion was that of argument by design, whereby nature was used as theological proof of god's existence. As noted by William Paley, nature's perfection reflected a designer, as Aristotle's chair and the craftsman who created it. Some typical examples were that of the human eye or the harmonious relation of bees to flowers. Ultimately, a reconciliatory consensus was reached in the US, claiming that there was no conflict between science and religion because the 'first cause' (God) would always be validated. Darwin even received honorary degrees and recognition in the United Sates a decade before England recognized his merit.

Social Darwinism's key postulates included the notion of 'survival of the fittest', a term actually coined by Spencer, as well as that of 'gradual change'. While Spencer played a key part in the formation of social Darwinism, it was of an indirect and unwelcome manner. He was enormously popular in the United States—a little bit too much so. There are humorous anecdotes regarding his unwelcome fame and popularity. Spencer was actually a rather shy individual who did not like public events

---

[53] The formation of reproductive cells occurs at very beginning of cellular division, which are then completely isolated from rest of the body. Thereby, even the external DNA from other cells in the same body do not affect future sperms and eggs (human reproductive cells.)

very much; but he was idealized by Andrew Carnegie, and that made all the difference. In fact, Spencer criticized the reception of his ideas in the United States, noting that the overemphasis on work was unhealthy for the body. His notions became grossly distorted in the US, and were used as a justification for the existing social order, and as a validation of the status quo.

Some social Darwinists in the United States included William Graham Sumner, a Yale theologian who had read about economic principles as young man. His father had been a typical middle class migrant to the US, characterized by a strong work ethic and frugality. Sumner attacked liberal writers as Upton Sinclair, author of *The Jungle,* arguing that social reforms could not quickly modify what had taken eons to develop.[54] Millionaires were millionaires because they were the best men suited for the job; since they could handle the difficulties of complex tasks, they consequently required higher remuneration. Sumner's views overidealized competition, creating stances which were repeated ad nauseum in the popular press.

Lester Ward countered that it was naïve for social Darwinists to believe that businessmen won by fair play; a great deal of unscrupulous behavior occurred in the world of commerce. He also pointed out that there was a great deal of inefficiency in the natural world. Fish laid millions of eggs, of whom only a few survived into adulthood. Contrary to the claims of the social Darwinists, nature was not an ideal model to follow. When taken to extremes, competition was paradoxically self-defeating as it typically resulted in a monopoly, and hence the elimination of the very mechanism by which social progress was established: competition. Commercial activity in a capitalist free market system required governmental intervention in order to retain its competitive character.

Religious ministers also attacked social Darwinism, aptly noting that excessive competition would not necessarily result in a civic state. The harsh world of competition would only create a land of brutes lacking sympathy for others in a community.

---

[54] *The Jungle* described labor and sanitation abuses in Chicago's meat industry.

Fortunately, men do have conscience, and the laws of nature should not necessarily be the laws of society.

The most potent attack against social Darwinism, however, came from the Pragmatists: William James and John Dewey. James came from wealthy class and had actually been an avid Spenserian in his early days, prior to the suffering of a mental breakdown and consequent nervous crises. Ultimately, he came to detest a philosophy which proclaimed to know the 'inevitable' outcomes of social formation. Dewey, the very same inventor of the library classification system, came from more humble beginnings. His father had been a small merchant, for whom frugality was important value. His *Ethics* (1908) noted that the historical origins of ethics did not constitute proof of validity of an ethical system, even if it is presumed that one could actually trace the biological origins of morality. In fact, one of the core principles of pragmatism is the notion that ideas are to be judged by their social consequences rather than their inherent truth content, from which it might be hard to establish.

Dewey was also perplexed by social Darwinism's deification and idealization of survivalism, as there was no notion of the ends or the purposes of such a social system; i.e. 'for what?' The inevitability of Spencerianism also meant that there was no room for alternatives. Dewey pointed out that all moral decisions were unique; each individual's circumstances are unique and different onto themselves, thus making universal moral judgments tenuous.[55] Circumstances are always complex, and hence it is hard to generalize across these. Ultimately, it is the individual who must decide for himself what the right course of action he is to take; to decide what as correct given their own unique and particular circumstances. Social Darwinism essentially ignored the existence of novelties in the historical process and denied the role played by cognition in human behavior. Individuals, Dewey points out, were not blind actors, but rather had consciences which shaped and dictated their own behavior. Because man operated in society, he was removed from the vicious calculus of natural selection.

---

[55] One never steps into same river twice, noted Heraclitus.

Some of these criticisms had already been expressed by biologists as Alfred Russell Wallace and Thomas H. Huxley. In spite of his enormous role in the rise of Darwinism, Wallace was oddly anti Darwinian. Wallace believed that natural selection applied only to the body and not to the brain. Cultural evolution proceeded distinctly from biological evolution, and at a much faster rate. Wallace also reflected the religious crises of the era. While is adoption of spiritualism and séances is today seen as an oddity, such spiritualism was a natural response to the moral crisis of period. Spiritualism filled a moral hole in the existing cultural realm, a phenomenon that is also observable in the social and religious history of Puerto Rico .

For his part, Huxley was also oddly anti-Darwinian, in spite of his ardent defense of Darwin in the public forum.[56] Unlike Darwin, Huxley was unwilling to draw any evolutionary conclusions that would be applicable to the social realm. In contrast to Wallace, Huxley was a much more experienced man, and had participated in Parliament and other leadership positions in England. As a result, he was much more sanguine about human behavior. Men were not saints, and was well aware of the social implications and the abuse such ideas could be put to. Huxley had also lost a son, which affected him deeply. In contrast to others, Huxley therefore drew a line in the sand in his essay "Evolution and Ethics" (1895). For him, man was in a constant conflict with the universe, seeking to place order to the entropy enveloping him. While conflict characterizes the universe, man naturally sought peace and stability. The notion that one should draw an ethics from the universe was to him absurd.

We may briefly conclude that the fall of religion due to the rise of science created a crisis of conscience that began to emerge during the eighteenth century. Christian religious moral claims which had been built upon a foundation of Greek 'science' came tumbling down, creating a moral vacuum which was believed could be filled by science. Early in this transition, natural theology harmoniously coexisted side by side with natural

---

[56] Huxley criticized himself for not thinking the idea after having read it.

history, but the two would be eventually torn asunder. Yet, while morality became divorced from theology, it could not quite be attached to science.

During the nineteenth century, social Darwinism was one such response to the moral and religious vacuum. Shaped by its interaction with US culture, it created a set of values which reinforced US economic creed of autonomy, capitalism, and self sufficiency; however, it did so at the cost of society itself. Society was defined to be a vicious grinding machine rather than a source of support traditionally taken to be. Many feared that the emotional bonds which allowed society to cohere would be broken and lead to its dissolution and social chaos.

Ultimately, however, what undid social Darwinism were the very same forces which had undone Catholicism as well. Social Darwinism committed the error of establishing a worldview based on scientific theories. Should the scientific facts change, the worldview on which it is based will be placed into question. Darwin himself feared that if he built evolution on the basis of geology, his biological theory would come tumbling down upon unforeseen changes in geological facts or theories. The fall of Lamarckism ultimately led to the collapse of social Darwinism, and the decline of its validity as a social theory and ideology.

However, in its throes a new social philosophy was created which was worse than its predecessor: eugenics.

Ironically, Weismann's study led to a philosophical shift that would result in eugenics, or the selective breeding of individuals. This was a type of artificial breeding akin to the breeding of dogs for particular traits, but applied to human societies. Sexual selection was Darwin's counterpart to natural selection, described in *Descent of Man*. In essence, the female selection of particular male traits not only leads to competition between males, but fomented the development of such traits. Because the cost of fertility is very high, females tend to be picky, in which turn places pressure on males of all species to demonstrate 'fitness' (physical or genetic). However in the case of eugenics, the selective mechanism is transferred from the female to the state, or specifically to those in positions of public authority. Here the decision is removed away from individual, leading to an obvious conflict between individual and social interests. Agency, under the regime, is entirely removed from the individual while conscience and personal preference is transferred to the state.

However, the intellectual shift to eugenics was gradual. In its initially form, eugenics was not only 'harmless' but 'positive' in that it sought only to promote increased fertility of good families. "Good" middle class families in competition with poor were often defined as being at a reproductive disadvantaged in that the poor tended to have much higher rates of reproduction. The emphasis here had been shifted from quality to quantity, and Darwin himself thought poorly of such kind of behavior by the lower classes.

Gradually, however, the eugenics stance became more aggressive, and turned 'negative' by focusing on the inhibition of reproduction by certain classes, social groups, and types of individuals. Those deemed unworthy to reproduce should be banned from doing so, a theme eloquently represented in the movie *Man of Steel* (2013). As Ka-el implied, such a system removed agency from the individual and eliminated the possibility of unique random and unforeseen combinations that were in greater harmony with existing conditions. Purposeful selection was biased towards dull conformity. "And who will determine whose blood lines will survive, Zod: you?"

Eugenics

There a few key differences between social Darwinism and eugenics, and it is important to make the distinction as the two are often confused for one another. Both occurred at roughly the similar time and were very prominent in the United States, even if less so in England. Both also help to set the context of Cornelius Rhoads's atrocities in Puerto Rico.

The two movements are characterized by different thinkers. Social Darwinism, as we have seen, was based on the ideas of Herbert Spencer, which in turn were shaped as these were developed in the United States by Charles Graham Sumner and others. The founders of eugenics are also a combination of British and North American thinkers. The original theoretical creators were Francis Galton, Darwin's cousin, and Karl Pearson, a mathematician. Its extension and development in the US occurred under biologists such as Charles Davenport, director of the Cold Springs Harbor laboratory, and Harry Laughling, Davenport's mentor.

The two were beset by different groups of critics. Those critical of social Darwinism were mainly philosophers: William James, brother of novelist Henry James, and John Dewey, known for his contributions to education and pedagogy. They also included Darwinists as T. H. Huxley and Alfred Russell Wallace. Eugenics, on the other hand, was criticized principally by a group of scientists who also happened to be close mutual friends. These included JBS Haldane, whom was once referred to as a force of nature by a French scientist; Lancelot Hogben; Julian Huxley, brother of novelists Aldous Huxley; and Herbert Jennings who helped UNESCO prepare an official statement on race.

The two social movements were also constituted by fundamentally different ideologies. The key lever of social Darwinism was the notion of 'survival of the fittest'. Social competition between individuals in a society produced the most fit, and over the long run the best society—which implied a laissez faire social policy with little intervention by the state. Failure, in theory, weeded out the bad. It was also characterized by a tendency to the idealization of businessmen; monetary and financial success were taken to be indication of inherent individual merit.

By contrast, the key to social progress for eugenics was obtained by the control of human sexual reproduction. According to the theory, germ plasm could be tainted by different racial stock, hence its concern regarding the differential reproductive rates by different economic classes and ethic social groups. While 'positive eugenics' sought to promote and aid good germ plasm, 'negative eugenics' sought to prevent 'degenerate' germ plasm from spreading. Success under its rubric was defined in terms of intelligence. In contrast to the former movement, eugenics was most certainly not typified by businessmen, but rather by the professional cultural classes: scientists, artists, musicians, and engineers. Eugenics drastically differed from its predecessor in that effective social policy implied an active interventionist state into the intimate life of its citizens, the complete opposite of laissez faire. Its potential for creating the most repressive tyrannies was typified by Nazi Germany, which was a 'logical realization of ideas' but in fact had taken the US eugenics movement as its model.

The two do share one commonality, however. Both social Darwinism and eugenics were discarded when the science upon which they were established became discredited.

The role of women was seen as particularly crucial in the eugenics movement. Her selection of a husband had enormous implications for society, as the selection of the diseased, unfit or undesirable mate would be ultimately harmful for the racial stock. The criteria of unworthiness was defined according to the standard criteria of the era: alcoholism, venereal disease, social welfare, poverty, and foreign nationality. Particularly noxious in this view were the Mediterranean from Spain or Italy, non-native speakers which typically had a low IQ when given exams under high duress.

The notion of an intelligence quotient was based on studies by Alfred Binet, whose principal purpose was the measurement of mental versus chronological age, or the maturity of an individual relative to that of his cohorts. One example might be that of a 6 year old child with the maturity of a 10 year old; or, inversely, a 10 year child whom acts like a 6 year old boy. These studies were modified by Lewis Terman at Stanford, who came up with the numerical IQ quotient we know to day, or the ratio of mental age to chronological age multiplied by 100. The two prior cases would be calculated as follow: 1) 10/6* 100= 166 (an elevated IQ) and 2) 6/10 * 100= 60 or a low IQ. The mentally retarded in institutions tended to have a low IQ, by definition. In spite that the eldest patient was 50 years old, they might have the mentality of a 12 year old child, and hence a low IQ. The measure was later modified by Henry Goddard into certain categories: 1) 'idiot' with a mental age of 12 years of age), 2) 'imbeciles' with a mental age of 3-7 years, and 3) 'morons' between ages of 8-12. These were not to be confused with the insane, who appeared normal but had diseased brains; at the time, a 'moron' was defined by cranial dwarfism. The focus, however, was principally on 'feeblemindedness', a rather vague term, which comes form the Greek work 'kakos' or 'bad'. Eugenists estimated that roughly 1-3% of the population was feebleminded.[57]

These notions seemed to be proven by the available evidence. An Ellis Island study showed that 2/5 of migrants to the United States were feebleminded, statistically demonstrating that Mediterranean migrants had particularly low IQs. Concern arose regarding their impact on native germ plasm stock, particularly in light of the propensity to violence by Italians. It was estimated that these changes led to an increase in taxes, in that crime now cost $500 to the regular US taxpayer of the early 20th century. The principal concern with this migration was the 'mongrelization,' or the mixing of the races, and the US inability to carry forward Western civilization.

---

[57] In Puerto Rico, the term *caco* refers to youth with a propensity towards criminal activity.

A case which appeared to provide valid evidence for these ideas was that of the Jukes family.[58] A study in 1877 encompassing over 600 individuals in this one particular family, showed that 300 of its members were identified as being 'antisocial' with consequent social problems: criminality, prostitution, and so forth. A follow up study in 1915 appeared to reinforce the original study's conclusion, where the Jukes appeared to retain strong anti-social traits over various generations. Other family studies were also done, tracing seven generations, reaching similar findings. These studies suggested that social traits ran in families, allegedly presenting proof of the role of germ plasm in human behavior. Morality and ethics was ultimately the byproduct of heredity, which in turn had enormous personal implications. The wrong marriages could have disastrous results, leading to a life of infamy, poverty, crime, prostitution, and all sorts of unimaginable social ills.

Since the dilemma of reproduction is beset by all, the wives of the scientists involved in the Eugenics movement provide good case studies for the above dilemmas. What was their actual practice? Pearson formed a Men and Women's Club, where he met his future wife, an independent woman. He assured her that he was only interest in her mind. Upon their engagement she suffers a nervous breakdown, fearing the loss of independence, and took six months to recuperate. They eventually get married, and have kids; she becomes a traditional mother. It is to be noted that Pearson was somewhat 'tyrannical', and easy to see why the emotional crisis occurred in the first place.[59]

---

[58] The name "Jukes" was a false family name.

[59] Given that a woman's key biological function is that of giving birth, there is a universal tragic tension in the choice between her role of professional career and cultural contribution versus her role as a mother. This tension can be very hard, if not impossible, to reconcile, given that one often comes at expense of other, akin to that seen in the movie *Sophie's Choice* where the protagonist is asked to select between her two children in a Nazi concentration camp. The choice is particularly tragic if woman has excelled in a field, demonstrating high ability and level of achievement. She operates under a biological time clock; children born to women over the age of 40 tend to sufferer from mongoloidism; her natural biological limit is around 54 years of age upon the onset of menopause. In this context, it is important to note that professional

For his part Jennings married a biologist, who became an active coworker in his laboratory. He claimed that monogamy was happiest state for men. Weldon, a biologist helping Pearson, also had active assistance from his wife who did tedious measurements of animals and helped to create data for much of Pearson's work. Hogden for his part married a social liberal philosopher. Although her initial goal was to have a parallel career to that of her husband as a philosopher, she also ended up having four children and assumed the role of a traditional mother. Davenport, the leader of eugenics in the US, married co-biologist Gertrude Cotty, but Cotty seemed to have had a dominant role in the relationship. She seems to have pushed her husband for higher roles, and might have influenced Davenport into the rental of his home on an adjacent piece of property to the Cold Spring Harbor, whose rooms were rented to students and visiting professors. These rentals provided additional funds for the couple.

In contrast to the traditional motherly role, the emergent use of contraception and birth control was actively adopted by the US feminist movement. Feminist Margaret Sanger argued that contraception should be given to poor, as it would slow down differentials in reproductive rates. The poor should also be given access to the same opportunities as the middle and upper classes, according to Sanger. There were a number of arguments, however, against the use of birth control—principally that it divided sexual activity from reproductive function. It was feared that this division would result in a greater focus on pleasure than on intimacy, which in turn would lead to an increase in licentious behavior and a gradual decline of morals in society.

While families as the Jukes were prone to immoral and unethical behavior according to eugenic theory, they also provided guides to social policy. A $150 sterilization of the original couple in the Jukes case would have saved the state $2

---

and academic productivity peaks vary enormously per field. While that of mathematics is young (24 years of age) given its dependence on mental acuteness, that of other fields as astronomy shifts to later years (40s), as it is dependant more on organizational skill.

million, showing that sterilization could be an important tool of social policy. Many other solutions were also proposed. It was argued that economic dependence forced women into bad unions, and hence that a female's financial insecurity inevitably led to marriage with undesirables given the absence of other alternatives. As a consequence, it was proposed that a college education for all women was deemed imperative, as it would lead to their financial independence, and thus allowing for greater selectivity of mates and husbands—with its consequent social implications.[60]

Theodore Roosevelt and others were concerned that the reproductive rate of middle to upper class women would fall. Four children per woman represented the ideal reproductive rate, which led to both the replacement of their parents and an additional two more; this was seen as contributing to a manageable growth. It was disconcerting when a study done of Harvard graduates over 25 years showed a replacement rate of only ½ to ¾. This meant that the 'good' were not reproducing at same rate as the less talented in society; which ultimately meant that the US germ plasm would degenerate over the long term.

In spite of their cultural similarities, there were distinct concerns in the United States and England, which lent each social movement their divergent characters. The US was pervaded by a concern with the rise of crime, prostitution, vagrancy, and a whole host of other social ills. England, on the other hand, was more worried about the decline of empire. It required the best members in order to compete with emerging global powers as Germany.

One might suppose that George Bernard Shaw's *Man and Superman* (1902) dealt strictly with male issues. He pointed out in his criticism of British colonial policy that it did not make sense to remove Tibetan or Lassan leaders from their land, if the colonial metropolis did not guarantee the superiority of the ruling Englishman. He was very interested in the eugenics movement, and did go out of way to participate. He met Pearson and others,

---

[60] As they did not need money, they would also not be forced into undesirable arrangements, which in turn created other macrosocial problems.

and became an active representant in England. However, the book should have been titled *Superman and Superwoman*, as a great deal of emphasis was placed on the woman's role in the good society—specifically notion of mate choice.

Shaw argues against limited mate choice circumscribed by an immediate social grouping. Typically women chose from men in close physical proximity; those in her similar class, social group, education, or geography. This was a grave mistake given that, by definition, her range of selection was limited. For Shaw, society needed to encourage marriages across social boundaries. By providing the most ample number of potential mates to a woman, she would then be able to select the most ideal candidate, and thus help push the British race onward and forward.[61] Simple observation of contemporary popular artists as Jennifer Lopez, however, provides a hint of the limitations of Shaw's proposals.[62]

All of this might suggest that eugenics helped spur feminism, which it obviously did not. Traditionally, the role of reproduction and family were seen as a woman's key domain of activity and control in society. The household included her family, the children, and all other maternal relations. However, the movement actually sought to transfer this domain from the private to the public, from the individual to the state. In this ambition, eugenics essentially sought to remove the woman's personal choice of mate selection to a superorganism (the state),

---

[61] Note here that the focus is on the social group, rather than cross racial breeding.

[62] Consider the following hypothetical example. A local group might consist of 150 persons (tribe), and crossing barriers might amplify this to 1 million. The choices available to the singer Jennifer Lopez will be incredibly wide ranging, as suggested by Shaw, in the millions due to the magic of modern media (radio, television, internet). Her range of mate choice is very high, and is perhaps why fame is so desirable; she has an ample selection to pick from, akin to the Home Depot versus a local hardware store. However, the problem resides in the drastic increase of mate choice. As the number of mates increases, its logistics become impossible and unmanageable. Jennifer Lopez simply lacks the time to look deeply into all of her potential mates, and hence selection will by definition again become superficial. In the real life, couples typically marry after 2.5 yrs of courtship, having enough time to know the personal traits of their partner.

turning its actual character into an 'antifeminist' one—for which it should not be mistaken.

How can we account for the origins of an ideology that was to have so much influence and impact during the twentieth century? It was incredibly popular and adhered to throughout world during the first two decades of the century, in England, the United States, Germany, and in other European states as well. It even spread throughout Latin America in countries as Brazil and Argentina, with substantive variants.[63]

Perceived social ills are the principal factor for its adoption in the United States. The rapid urbanization, originating in both immigration from the rural countryside and from Europe. Between 1900 and 1915, half of the total US population was made up of migrants—a rather monumental change in demographics, which in turn modified the social layout of the land. In a relatively short span of time, an individual would not know many of his neighbors, for example.

By contrast, the principal factor in the adoption of eugenics in the United Kingdom was the concern for the degradation of empire; the nineteenth century is known as Britain's century given her dominion during much of the world at the time. However, by the late nineteenth century, she had strong competition with the ascendant industrial Germany, who forced the partition of Africa, and much of the rest of world. The predominant concern was that a degradation of the quality of British 'stock' would hinder empire. Eugenics, however, never became as dominant in England as it did in the United Sates.

There were also disciplinary factors in the rise of the ideology. Momentous changes had occurred in philosophy, whereby psychology, previously a branch of philosophy, separated itself from her mother discipline. The pragmatist William James came to be seen more as a psychologist, and the field was attempting to make itself more 'scientific'—hopefully with greater respectability. Some of its key leaders had been humiliatingly denied academic recognition, as Robert Keyres

---

[63] It goes without saying that, given the multiplicity of variants and geographical locations, the subject is complex, at both a personal and a social level.

who came up with new Binet test and was President of the American Psychological Association. Keyres was denied tenure at Harvard University in spite of his academic achievement and recognition.

Similarly we may point out that the 'new biology' was coming into its own. It wanted to draw more on physics and chemistry: divide objects into its component parts. Traditional biology was typified by natural history, with its holistic, descriptive, and comparative character. Changes in biology, specifically the rise of genetics, ultimately continued to shift the underlying scientific tectonic plates on which eugenics existed, discrediting it at turning it into an empty ideology without substantive scientific underpinning.

Eeugenics emerges from the work of two individuals from England: Francis Galton and Karl Pearson. Interestingly, their personal histories shed some light on the formation and character of eugenics, even if the final movement was not theirs in the long run.[64]

Francis Galton (1822-1911) was born into an upper-middle class professional family and had shown signs of early promise as an infant. His sister Susan was born with a curvature of the spine; being forced indoors, she doted a great deal of energy on her younger brother. Galton could read by the age of two, write by 4, and read Latin by the age of 8. He excelled in mathematics, and by the time he went to college there were very high expectations with regards to his future achievement. However, when taking the famous Tropos exam in mathematics, Galton came in second

---

[64] Again, eugenics was greatly modified upon its arrival in the US by Davenport and Lilligan, whom were both were criticized by Pearson for their shoddy research. A notion from critical literary theory is rather apt: an author only controls work when in hands. Once set to the world, the work obtains its own dynamic, and the author now cannot control how it is interpreted. In the history of technology this process is referred to by the dynamic of unintended consequences, akin to Frankenstein of sorts. One logical consequence is that we cannot place full blame on both Galton and Pearson, but rather might refer to dynamic as a type of 'perfect storm': the right combination of factors led to unexpected consequences.

place in spite of his immense effort, which to him was utter defeat and sparked a nervous breakdown.

After Galton withdrew from school and his father died, he inherited a sizeable fortune, whose magnitude meant that he no longer had to work to earn a living. So, the young Galton decided to go on trip of Egypt with some of his friends. The typical British trip to Egypt consisted of sailing up Nile, which he did, to the first cataract. However, here Galton meets a French intellectual "Arnaud", who lived a simple lifestyle in a hut with scientific instruments, including a thermometer, barometer and a few scientific works. Arnaud served the local sultan, and challenged Galton to explore Egypt beyond the Nile.

Galton did this, crossing the dessert and going to Khartoum. This journey was an important symbolic step, both literally and metaphorically for the young man; it was a coming of age and Galton recognized that his life distinctly changed afterwards. He became an archetype for the likes of Laurence of Arabia, not only by interacting with locals, but in heroically solving local disputes.[65] Galton also explored South Africa, which at the time was poorly known. An incident reveals his temperament. On one occasion his boat overturns, and he is taken 200 yards underwater. Galton, with a strong athletic build, could remain calm and cool under pressure. His scientific work can be characterized by its dashing character; Galton often jumped into unexplored fields in science, as if in a jungle or desert.

Galton also loved to count, and applied this counting widely. For example, it is he who established fingerprint as crime tool, as the probability of any two fingerprints being exactly alike is very low. He also applied his love on numbers to meteorology. While his lines of similar temperature and pressure were not original, they did further the field. That being said, there is not doubt that his application of statistics to biology was an original contribution, and became particularly important in shifting the character of biology. Until Galton, nobody studied biology in this manner. He recognized that complex data could be given order

---

[65] At one point he rode an ox to the front door of a tribal leader, making a successful show of force.

and coherence via use of statistics—even if it was a somewhat shallow analysis that never peered into the internal mechanism of a system.

One final aspect of his life is important to the eventual character of eugenics: Galton married but was never able to have children, which personally tormented him—particularly the notion of original sin. It appears that he did not conduct himself too gentlemanly during his African adventures, perhaps typical of young colonizers. He acquired a venereal disease, which might have caused his sterility. Galton's inability to produce children in spite of his ardent desire to do so was accounted for by him to his personal sins, which in turn created another significant personal crises. Over the course of his life, Galton actually had several nervous breakdowns. In one of these, his discovery of Darwin's *Origin of Species* was akin to a religious revelation for it destroyed the notion that men had descended from a prior state of perfection. The history of man, Galton learns, was one of gradual ascent, which would continue indefinitely onto the future. It goes without saying that such notions personally afforded Galton a great deal of hope for future, which he had previously lost. He would try to shape the future of all children.[66]

Galton began by first studying the *Dictionary of Men of Time* which contained biographies of distinguished men of the era, typically from the upper echelon of society. He traced all of their family histories and found that brilliance tended to run in families. It goes without saying that there were many errors in his analysis. Principally, he presumed that social reputation equated brilliance, but in fact not all famous persons are smart. He also ignored the role of social context. Brilliant man afflicted by a negative environment confront too numerous obstacles to success. Note as well that the upper class did not typically enter the public domain. The mentally retarded from the upper strata were usually sent to private institutions, which in turn kept them out of the public record and led to a false upward bias in his study.

---

[66] Darwinism appears to have been projected by him onto the rest of society.

It should be mentioned that the broader statistical study of society at the time was afflicted by many limitations. 'Statistics' typically meant that some sort of social data had been gathered but which tended to be limited in quantity, in either range or size. It was also limited in quality in that usually there was no mathematical analysis or theoretical evaluation undertaken of such data. Galton likened his efforts to those of the Israelites of Egypt, whom not only had to make bricks but to construct the very materials from which such bricks were made.

Galton remedied his absence of data in two ways. He institutionalized his efforts by forming the Anthropological Laboratory in 1884 at the International Health Exhibition of the Science Museum. In it, he offered 500£ for the complete data of an entire family, and in the process managed to gather 9,000 family samples. He also initiated a study of sweet peas, growing many, and sending these to friends throughout England, including Darwin himself. These returned the peas to Galton, who thereupon performed statistical analysis of his samples.[67]

One of Galton's most important contributions was creating the innovative *coefficient of correlation of parts*. Before him, the Gaussian curve was typically used as a measure of error and as a means to obtain the 'correct number' in physics. Galton redefined it as the distribution of traits in a population. Ironically, his early findings actually pointed to conclusions opposite to those of eugenics. When he compared only two generations, rather than identify a cascading trail of evolutionary change, he noted that their traits tended towards a same distribution, or to a mean. He interpreted such curves to represent the 'pull of ancestry', a puzzle he could not solve. In other words, the 'father of eugenics' actually disproved notion that traits could be improved over time by breeding in his earliest studies.

---

[67] It is curious to note that Galton was born in same year as Mendel. Most wrongly assume that Gregor Mendel was a monk by temperament. In fact, his training was not in theology but rather sciences. The Brunn monastery wanted to improve its agricultural output, and hence is stimulated to brings him in as a monk. We might account for the lack of reception of his work in 1866, following Darwin's in 1859, on the basis that biology then was focused on change whereas Mendel's work had been on stability.

It is important to note that, in spite of his mathematical ability, when in a tough spot, Galton tended to give the work to professional mathematicians—which is where Pearson steps in. It is Pearson, and not Galton, who proves that Galton's early notions were 'right' . . . sort of.

To first understand Karl Pearson (1857-1936), however, we have to look at Walter Weldon. Both immediately understood that the application of Galton's statistics to biology ushered a vast improvement in the field. While Pearson was a mathematician, Weldon was a biologist comparing crabs in Plymouth Bay (United States) and the Bay of Naples in Italy. He studied 22 traits, nearly all whom showed regular mean distribution curves for the two species. However one particular trait, their frontal breadth, remarkably showed a dual bell curve. Weldon along with Pearson immediately recognized its importance for biology: statistics could provide a non ambiguous and exact definition of a species. Previously morphologists spent a great deal of time in endless debate on the adaptive significance of a particular feature, which was then used to pinpoint a particular species. These intense debates, however, tended to be highly speculative, and thus were not easily clarified. Weldon recognized that the application Galton's innovation provided an unquestionable criteria for the identification animals that on superficial inspection appeared to belong the same species. Both Pearson and Weldon were eventually inducted into Royal Society of London for their contribution.[68]

Pearson realized that Galton was onto something and sought to extend it. His first studies were undertaken of intelligence in British schools when IQ test had not yet been invented. In the system, teachers assigned discontinuous traits or discrete variables such as 'dull', 'slow learner', 'quick intelligence' to each student. To understand the role of heredity, ideally one needed parental scores and their own teacher evaluations, as well, but such long-term records were not kept. As a result, Pearson created an innovative substitute: he correlated data between

---

[68] Weldon seldom published, and much of his research was actually incorporated into Pearson's numerous publications.

siblings to see if their respective evaluations matched, as a proxy of relatedness. To undermine the potential intrusion of a social bias, he made a physical correlation as well as these would not have any direct social determinants. He appeared to have proven the hereditability of intelligence, and set off on a lifelong career in the topic. As with Galton, Pearson's biography is of interest.

He was often defined as an 'ice-cold' person whom seldom warmed up to those around him—incidentally the psychological opposite of Weldon. Pearson tended to attack any who criticized him, even when such observations were legitimate. His father had been a successful lawyer who served as counsel to the queen, whose busy schedule tended to completely remove him from the family setting.[69] Pearson's mother was horrifically afflicted and miserable, which made her son acutely aware of the detrimental effects of women's financial dependence. It is Pearson who publicly argued for a woman's financial freedom, likely in light of his personal background. A position adopted by Shaw, he believed education should be afforded to women so as to not limit mate selection, and which should cut across social circles to extend outside of a woman's own immediate group.

Pearson as a boy also did not get along with other school kids. In contrast to his elder brother, he was not sent to a good school but rather to one of lower classes, wherein his peer's attitude to education was a hostile one. Naturally, he tended to isolate himself completely from other students, focusing on his school work, and did quite well.

Upon his successful entry to college in Germany, Pearson changes his name from 'Carl' to 'Karl', and was very influenced by philosophical trends at time, which included Kantian idealism, Goethe's romanticism, and Comtian positivism. The last doctrine claimed that science could only be based on tangible and testable evidence. As there could be no invisible entities, its influence might account for Pearson's actual hatred of genetics and eugenic popularizers who tended to base their claims on Mendelian elements—which at the time could not be seen or detected and

---

[69] It is said that he quickly downed breakfast, only to return very late in the day.

hence was defined by Pearson as deficient. For Pearson, 'genes', or what were then called "Mendelian elements", had no scientific validity as they were only artificial hypothetical constructs.[70] Pearson came to define his science as 'biometry' or the statistical analysis of measurable traits. It goes without saying that his 'eugenics' was not the same as genetics, strictly speaking.

Unlike Galton, Pearson did not have a family fortune to prop him up, so he tended to routinely shift from one academic job to the next. The emerging professionalization in science had a positive effect in that it forced practitioners to operate at high level. Its detrimental aspects included huge teaching loads, which detrimentally affected research. For his last class, he had to grade 200 final exams. Pearson became friends with Galton, which was of a greater boon than Pearson could have expected. Galton passed down his sizable inheritance to Pearson by making him director of the Galton Laboratory for National Eugenics, previously called the Eugenics Records Office. As its director, Pearson also began journal the *Biometrica*, whose articles were mainly results from his laboratory's findings, and began a series on "Studies in National Degradation". Again, Pearson's focus was not on individual competition but rather the international one of the British Empire. Darwin himself had once expressed his concern to Wallace with regard to the implications of the differential rates of reproduction between classes, as these would be ultimately detrimental not only to England but to humanity as whole—a concern shared by the novelist H. G. Wells.[71]

Towards the end of Galton's life, Pearson insisted that Galton offer lectures to the general public about his work, and in fact ended up becoming an avid lecturer. Galton drew large audiences with his multifarious stories and adventurous tales. Galton was knighted by the Queen and also obtained the Copley Medal given

---

[70] Darwin proposed the idea that traits passed by 'gemules' in the blood. Galton actually tested for this idea by injecting white rabbit blood to grey rabbits, but the latter did not alter color in the next generation, and hence disproved Darwin's early 'genetic' notions.

[71] These issues are discussed in novel *Time Machine*. It is known that Wells had been an avid eugenist who attended lectures by Galton, even if he later came to disagree with it.

to him by the Royal Society, its highest distinction possible. These lectures likely began the trend in the popularity of eugenics. However, when the field arrived in the US, it became severely distorted.

The enormous demands generated as a result of the First World War meant that any tool assisting in the selection and sorting of soldiers would receive a great boost. Under these conditions Robert Keyres had been given a tremendous opportunity by the National Research Council; he was instructed to create a test for mass application to new recruits, so as to efficiently determine their intelligence and classify them accordingly. Keyres established two forms, the alpha and the beta test. The first was a typical assessment testing for verbal skill, reasoning, and so forth. The beta test, on the other hand, was a pictorial test for the many illiterate recruits, given to men who had never held a pen or pencil before. The assessment was successful and helped the army place millions of men in appropriate positions, relative to merit. The results, however, were classified as a national secret and were not revealed until after the war had ended. The officers in charge of the new recruits detested the exam because they felt they had enough experience to personally determine the quality of a man; it goes without saying that the exam removed their own agency in these affairs.

When the report was finally released to the public in 1921, its results were alarming. The *Psychological Examining in the United States Army* found that average draftee had 72 IQ points and that 25% of these were unable to read. The shocking results were made public, and suggested that the warnings of a national decline had been valid after all. Incidentally, similar findings had also been made in England. Children were entering mental asylums at twice the rate than colleges, and the number of criminals had substantially increased overall. While these exams were highly criticized by the likes of Herbert Hoover, who claimed that the results were due to poor nutrition, he was forced to retract his original statements, given that they contravened 'all scientific knowledge of heredity'. The 11th volume of the *Encyclopedia Britannica* stated that the "the organic betterment of race [occurred] through the wise application of the laws of heredity."

The results implied the need of a drastic policy change; both positive and negative eugenics would be needed to allay the negative national trends. Havelock Ellis criticized that without the sterilization of the feebleminded, all efforts towards national progress amounted to a futile Sisyphean task. It is somewhat shocking, from our point of view, the number of eugenic measures that were passed in those years. California was by far the most active state, undertaking 6,244 sterilizations by 1929, or twice that of all other states combined. The total in the US equaled nearly 36,000 sterilization by 1941, most of these occurring in state institutions. These included vasectomies for prisoners that were undertaken on the slightest of pretexts. Families on welfare were hunted down in rural areas of Virginia, which in turn had the second highest rates of sterilizations in the nation.

The radically overbearing governmental activism which such policies implied ultimately ended up in the Supreme Court under the case of *Buck v. Bell* of 1927. The case involved Emma Buck, an institutionalized feebleminded mother; she had one child prior to her institutionalization, Vivian, and another girl afterwards, Carrie. At issue was the forced sterilization of her children. Harry Laughling of the Cold Springer Harbor institute provided 'scientific proof' that both children were 'feebleminded', thus giving alleged scientific credibility even though he had never met Vivian, who was actually bright. Tragically, the case passed nearly unanimously in US Supreme Court, and votes in favor of the measure included liberals as Justice Louis Brandeis, an anti corporate justice. The final opinion was even written by renown Justine Oliver Wendell Holmes. Only Justice Pearce Butler, a conservative, opposed the ruling, but did not express his reason, likely being its blatant and overbearing incursion on personal liberties. Wendell Holmes explained that the court wanted to prevent the nation from being swamped by incompetence.

Were the eugenicists correct in their assertions? Are these issues only factual questions whose only criteria should be strictly scientific or should we adopt William James' position and measure claims by their social consequences?

There can be no doubt as to the foreseeable ends and consequences of such policies. Walter Lipman was indignant as to its pretenses. " I hate the sense of superiority which it creates, and the sense of inferiority which it imposes." While their observations with regard to the statistical increase in crime may be accepted as valid, their causal explanations were off the mark. New urban studies by physicist Geoffrey West demonstrate a distinct correlation between population density, crime, and patents. As the log of population density increases, there is a directly proportional increase in the amount of crime. However, a correlation can also be observed with regard to many other traits, such as the number of technological patents, the number of gasoline stations, and so forth. It turns out that cities are dual edged swords, in that they allow for a much richer interaction of individuals, and greater intellectual creativity, while at the same time produce noxious side effects as crime. Both technological patents and crime increase proportionally relative to the population density of a region. In other words, we cannot have the good (innovation), without also having the bad (crime); for the rose to exist, it must have its thorns.

There were many problems with Pearson's work. In spite of its contributions, there were core errors which were hidden by sophisticated mathematics. His correlational analyses hid implicit presumptions. For the true identification of a dependent variable, other variables need to be held constant, as seen in bullet speed tests. A bullet's speed is influenced by the shape of bullet, temperature, the density of air, and so forth. Each of these variables can be modified or held constant while testing the other ones to determine their relative 'correlation coefficient'. However, such an analysis of intelligence is impossible given the large number of uncontrollable variables.[72]

---

[72] Genius twins in different environments would represent 'ideal studies' as hypothetical William Shakespeare, Benjamin Franklin or Louis Pasteur twins from lower classes. Such studies were undertaken by Freeman/Holzinger who identified only 19 twins in various foster homes after decades of searching. Their principal finding was of a close correlation between environment and

Most of eugenics's problems, however, were principally scientific, as noted by the biologist JBS Haldane. The notion that a 'race' is a genetic entity is a false claim. Jews, for example, exist across many different population groups, and thus are marked by an enormous amount of genetic variation. There is no single 'Jewish' genetic trait properly speaking, and do not represent a truly distinct genetic group. Haldane made an ironic quip about Hitler and his staff. It was odd that they praised blonde blue-eyed Nordic models, when none of them fit according to this standard. Hitler had black hair, Herman Goering was extremely fat, and Joseph Goebbels on the other had was a skinny fellow. Race was incorrectly being used to define ethnic group, when in fact it was characterized by a substantial amount of genetic variability.

Critically, the sheer notion of genetic perfection was deeply flawed from the standpoint of Darwinian evolution. According to Darwin, perfection is relative to environment in which an animal lives; more specifically a creature's level of adaptability relative to its circumstances. Haldane expressed this in press conference during Third International Eugenics Conference. "Describe heaven, and I will give you the saint." Contrary to the implied presumption, a homogenous population of perfect men would actually have a low resilience and longevity in that only by having a large amount of variability are populations able to survive drastically changing circumstances. As a consequence, it is not by establishing ideal forms, but rather by promoting diversity that human communities were strengthened.

Scientific courage was certainly important for society and science, at least in England. In contrast to the defense made by Haldane and his colleagues, nobody stood up to Trofim Lysenko in the USSR.[73]

---

intelligence. Sensitive intelligence cannot be fostered in environments dominated by blunted feelings, low cunning, and vicious propensities.

[73] Lysenko proclaimed absurd Lamarckian theories for agricultural production, alleging he could turn spring crops into winter crops. The absence of a serious challenge meant that Lysenko ended up killing 10 million Russian citizens when his policies resulted in the Great Famine of 1932. Even worse, the discipline of genetics in Russia was literally wiped off the map as a result of

Lysenko's close alliance with Stalin. More geneticists were killed in Russia during this period than physicist suffering in the Nazi regime. The most tragic aspect of all was the assassination of Nikolia Vavilov, who had given Lysenko his first important job.

Comparative and Legal Aspects of Bioethics

According to Leonides Santos Vargas, the father of bioethics in Puerto Rico, the word is a relatively recent term, meaning "the study of the moral and social implications of new technologies that arise from advances in the biological sciences."[74] The term was term first coined in 1970 by Van Rensselaer Potter. It is certainly the case that historical factors came into play in its emergence. World War II showed the extremes of evil to which men were capable. The systematic and institutionalized torture of other human beings by the Nazi regime raised questions with regard to the possibility of systematic evil on an unheard of scale in human history. How inhumane could humanity become via its use of biology?

When considered in light of WWII's scientific legacy, specifically the enormous power of science as demonstrated in the atomic bombs which destroyed entire cities of Hiroshima and Nagasaki, science had new and ominous implications. By the mid twentieth century the character and nature of science had clearly changed. After the war, the scientifically driven military industrial complex created devices with even greater power: hydrogen bombs whose megaton explosions could wipe humanity off the face of the Earth. Science was no longer the result of a lone genius working determinedly in his office, but a collective effort with enormous repercussions and social impact. When both are combined, the power of science and a ruthless government, its outcome could be horrific. What if the Hitler

---

[74] The original Spanish version reads, "*el estudio de las implicaciones morales y sociales de las tecnologías que resultan de los avances de las ciencias biológicas.*"

regime had actually developed a nuclear capacity?[75] Science now required an ethics, the absence of which was short-sighted.[76]

The rise of social movements in the United States also played a role in its emergence, including environmentalism, civil rights and feminism during the 1960s. Prior to 1950, automobiles had not been common in the US; however, the rapid increase during the century clearly led to pollution on a massive scale, as images of New York City prior to the *Clear Air Act* of 1963 attest. Rachael Carson's *Silent Spring* (1962) was also a bombshell. Multinational chemical companies improperly discarded their waste products in rivers and in the open. When leached into incubated egg shells, they would be unable to sustain the weight the of parents. The spring was silent because all the birds had died as a result—a form of genocide on a smaller scale.

Numerous leftist movements arose to meet these challenges, and had a broad cultural influence. Through collective social action, legislation could be enacted which would create tangible benefits to the community. Man made threats, unintended or not, could be controlled through sustained collective action. Women, whose participation in the labor force skyrocketed during the century, would now be given equal treatment; Pearson's dream had been realized. The rise of the corporation presented challenges continuously. Regardless of any entity's source of income, the greater the economic power, the greater its abusive influence in politics.

Hence revolutionary changes in biology, specifically the rise of genetics and the second Darwinian revolution, were placed in a new context; if physics could lead to the destruction of humanity, was biology any different? The ability of man to directly tamper with natural forms concerned many.

---

[75] At the time, the concern was not unrealistic as German scientists had been the fathers of atomic physics; Werner Heisenberg, who became an officer within the Nazi regime, invented quantum mechanics.

[76] Cold War Stalin engineers did not study ethics, which explains why hacking is so common in Russia as these are willing to do anything regardless of its costs. By contrast, all engineers in the United States have to take ethics courses.

In evolution, nature only modifies prior biological material, and thus has a limited range of action with regard to the degree of possible change. All 'natural' genetic change is cumulative, built on precedent; nature cannot erase and start all over from scratch.[77] However, the new science of genetics gives an external agent (man) an unprecedented amount of new power: the fine-grained and detailed manipulation of the genotype of all the animals on the earth. With it, biological changes are no longer constrained by evolution as it previously had been; man made irreversible changes could spill over throughout the biota. Philosophers became focused on the use and abuse of new science, and the concern was well reflected in both literature and cinema, as presciently seen in the novel *The Island of Dr. Moreau* by H.G. Wells. The new bioethicists were quick to note that their realm of study should not be strictly delimited to medicine, but covered the totality of biology; epoch making changes in the natural realm should not be left only in the hands of scientists.

We should not be fooled by the new lexicon, however. Ethical debates and concerns have occurred all throughout history regarding experiments in both biology and medicine. The oldest debates stretch back to the Hellenistic period, between the dogmatists and the rationalists. For over 2,000 years, there have been 'bioethical' debates, particularly so with regard to animal experimentation and so, strictly speaking, 'bioethics' is not an entirely new field. Through artificial selection, man has been modifying genetic code for centuries. The fruits and vegetables we eat today do not have the same shape and form as that of their ancestors. Watermelons today are more 'watermelony', being larger, with more edible material and smaller number of seeds as a result of man-made artificial selection.[78]

The debates were keenly felt in England, whose thinkers wished to extend the rights of man to animals. For John Locke,

---

[77] It does so to a degree over millions of years with each new geological epoch, but the evolutionary dynamics still apply.
[78] What is unique today, however, is that the new technologies of genetics allow for a much greater range of change and a wider ambit of unintended consequences.

the differences between men and animals were only one of degree rather than kind, given the role which senses played in our formation. David Hume argued that sympathy, rather than reason, was the basis for morality. Because animals also had senses and could perceive the world, these had to be included within any ethical-legal structure. Jeremy Bentham took a similar position. The father of utilitarism noted that a horse or dog is more rational than baby the day it is born; morality cannot be determined by the number of legs. As with Hume, he argued that morality was not based on the question of whether an animal could or could not reason, but rather whether it could feel. For him, the basis of ethics towards animals was principally based on sensation. If animals had the capacity to feel pain, then moral guarantees applied to them as well.

These views might be contrasted to those which emerged in France. Rene Descartes argued that animals could not reason; they had no mind, and were more akin to machines like a clock. While it has been argued the Cartesian stance was used as a justification for the abuse of animals, this is incorrect. The tradition of animal experimentation had occurred over a much longer period in France. Its experimental emphasis on the functional study of the body, or the 'how' life works (physiology) had also been greatly influenced by creation and rise of chemistry in that nation. Hence the prominent animal experimentalism of Claude Bernard during the nineteenth century actually goes much farther back in history to the time of the Parisian anatomists during the seventeenth century, which helped establish basis of animal experimentation.

These philosophies led to distinct legal changes. It was clear that there were asymmetries of power between actors. The scientific experimenter had absolute power, akin to a tyrant, in that he could do anything in private. The animal, by contrast, was totally helpless, akin to that of a baby. Animal experiments were thus typically never performed in public, as it would often cause public revulsion, outcry, and opposition.

Legal efforts to control animal experimentation first emerged in 1825 and were led by Richard Martin after witnessing Francois Magendie's merciless experiments in England. Magendie had been rather unsympathetic to the animal subjects used in a public demonstration, commenting for example that the dog experimented on would not move so much if it had spoken French. Unfortunately, these early legislative efforts were unsuccessful, likely because they were too innovative.

The principles of contemporary laws regarding animal experimentation were first established by Marshall Hall in 1830. These included the notion that experiments had to be absolutely necessary; all alternatives had been explored and evaluated prior to undertaking the final experiment. The experimenter needed to have clear and attainable objectives, thereby avoiding unwarranted and needless repetition of the experiment. A witness should always be present in these procedures, and the least amount of pain should be inflicted on the subject, promoting the use of some type of anesthetic.[79]

When the Cruelty to Animals Act was finally passed in 1876, fifty years after first proposals, it incorporated many of Hall's principles. The legislation had been lobbied for by the Society for the Prevention of Cruelty to Animals, headed by the wealthy activist and journalist Frances Cobbes. It required the licensing for animal experiments and routine inspections of such facilities. Cobbes, however, personally objected to the final law in that she wanted to abolish animal experimentation altogether. Unfortunately, her organization fell out of her control and shifted position towards a more moderate stance. Although Cobbes then formed a new organization, the Union for the Abolition of Vivisection, which pushed for the total abolition of animal experimentation. But it was not successful. Its ambition was perceived as being too 'extremist', falling outside the mean distribution bell curve.[80]

---

[79] It is curious to note that all surgeries, by definition, constitute a type of vivisection.

[80] It is to been noted that when groups take extremist views, they help shift the median viewpoint to that extreme. Whatever the outcome, these claims certainly recontextualize debate, or what is referred to as 'framing'. Extremes no longer perceived as extremes, a good example of which is the computer industry. During the 1990s, monitors of 15 inches were the norm, but are now perceived as 'small' relative to the much larger 30[+] inch monitors that exist.

There are three principles underlying the current legal landscape of animal experimentation: replacement, reduction, and refinement.[81] It is recognized that animals are essential for scientific research, but that such activities should seek to comply with certain moral standards, the key principle of which is to avoid the unnecessary suffering of any animal experimented on. One should see if an alternative experiment can be used to answer the research question at hand (replacement), as the case of in-vitro procedures. One should also use only the minimal number of animals needed to answer a specific question at hand; one should afford our fellow creatures an amount of dignity, avoiding the heedless and warrantless use of other living creatures (reduce). Finally one should modify procedures so that pain and suffering is averted as much as possible (refine).

Laws pervading the current landscape were established in 1985, some 160 years after the first efforts were undertaken. The Animal Welfare Act of 1966 was originally meant to prevent the theft of domestic animals (dogs, cats), but was extended to those animals undergoing some type of experimentation to insure their humane treatment. The Animal Welfare Act has been modified a number of times since its first draft: 1970, 1976, 1985, and 1980. By 1985, the Food Security Act, subsection Animal Welfare Act, was passed which provided a more precise criteria of evaluation and also required veterinarian care. It was criticized by animal welfare groups in that it excluded some of the creatures that are most experimented on in biology, specifically rats and mice.

The Health Research Extension Act of 1985, under the US Health Services, was complementary to the Animal Welfare Act, and is meant as an extension rather than its substitution. Its key features include the proper care and treatment of animal subjects. The participants have to submit an Animal Welfare Assurance documentation, whereby the number of animals, treatments, and procedures are clearly identified. As of this writing (2018), these reports can be downloaded from the US Department of Agriculture website. There is some question as to its effectiveness given that the reports are entirely voluntary.

---

[81] In Spanish, *reemplazar, reducir, refinar.*

Differences also arise with regard to the estimation of 'what to count', thus lending for a certain amount of variation and ambiguity in the reports. It is estimated that 96% of US research is conducted on mice.

Current legislation also requires the establishment of institutional Animal Care and Use Committees (ACUC). Any institution that receives federal funding needs to establish oversight committees over the organizations which perform them. The committees need to be composed of a scientist with prior animal experience, a relevant non scientific member (ethicist, lawyer etc), and a community member as well.[82] These committees in turn have a number of tasks, the key purpose of which is to provide an overview of institutional policy and the implementation of said policy. These reviews are supposed to be undertaken six months, but are often extended to yearly reviews. While reports of conduct and recommendations are prepared, ACUC bodies do have the power to stop research when gross violations are detected.

However, various problems have been identified in their implementation, reflecting typical concerns common to all regulatory bodies. Inevitably conflicts of interest tend to arise, and have been seen in agencies at the federal level, specifically the FDA (Food and Drug Administration) and the FCC (Federal Communications Commission). All too often a revolving door is established between the regulated actors and regulators; individuals who regulate are drawn from the industry that is regulated or turn to that industry for employment after their tenures have ended. Their high rate of turnover also implies that there is a lack of enforcement and consistency in policy. Unstable institutions are characterized by the loss of historical memory and consistent social action. Universities also have few incentives to comply with their ACUC boards. As universities can file patents on their internal employee's discoveries, generating sources of income in the order of millions of dollars, hidden roadblocks often emerge, as described by Marcia Angell.

---

[82] Its composition is incidentally like that of the board of Puerto Rico's *Fideicomiso de la Ciencia y Tecnología.*

The publicity of unwarranted animal experiments, however, can bring to the institution and its experimenters unwanted attention, and in turn to public attacks by overzealous animal activists, and the general loss of public support for such research. In a democracy, this loss of emotional support can have an enormous economic impact due to the reduced funding that might follow such potential outcries.

Some successful voluntary forms of compliance exist, as the AAACLAC (Association for Assessment and Accreditation of Laboratory Animal Care International) which was formed in 1965. A private non profit institution, it provides internationally recognized certificates of humane treatment in experimentation. Some 1,000 research institutions in 46 countries have been certified by the group, include the Sloan Kettering Cancer Center, St. Jude Children's Research Hospital, and the NIH (National Institutes of Health). Its standards were actually established by the US National Research Council, and are encoded in the *Guide for the Care and Use of Laboratory Animals* (2011), which can be freely downloaded from its website. Its standards go above and beyond those required by law.

We may make a number of observations. The underlying core presumptions of the current landscape include, but are not limited, to the following points: 1) animals are necessary for research and 2) a degree of moral responsibility to such animals is recognized. Given that the scientist must get on with his research, he must also so do in the least injurious manner possible. We might raise some question about these underlying presumptions.

In his critique of indigenous modes of behavior and morality, Thomas Hobbes noted that pirates, or any other thieves, typically have a 'moral code'. They will rob, steal, and cheat, but will try to minimize the danger to their victims. Pirate codes allegedly tried to limit the number of warrantless deaths incurred in an act of privacy. In spite of such generous concessions, Hobbes noted that pirates were still depriving their victims of personal property, hard work, or savings. As a result, there was an inherent unethical quality to this 'moral code', as all thievery is by definition a violent act.[83] Hobbes noted that both pirate and indigenous morality was a rather shallow one, and hence why they existed in a primitive state.[84] Locke himself established the notion of property rights as being fundamental human rights. Private property was inalienable, an extension of body, as it was commonly a product of hand, and hence as much a part of an individual as their very hand.

We may then consider the following hypothetical question. If I need to kill you (a human being) for my research , does my moral standing increase in any way just because I do not make you suffer?

While obviously an extreme hypothetical question with regard to men, in practice this is what has actually been done to countless of animals over the course of history. Would the Nazis be regarded with higher moral ground if they had not induced pain on their victims of genocide, some of whom were 6 million Jews? Would it have been any less horrific? The answer is fairly obvious; it is still horrific however we might try to rationalize the stance. A focus on reducing an animal's suffering, as contemporary bioethical legislation does, is somewhat shallow.[85]

---

[83] Nobody in their right state of mind wants to be dispossessed of their own private property. Similarly the arbitrary confiscation by the state constituted legalized theft; even if legal, it still remains unethical. These actions were all too common during the Colonial period and are often seen in Third World countries.

[84] As noted before, under this morality, society was 'brutish and short', or a war of all against all others and one that ultimately required a strong state.

[85] Such a stance is tantamount of depriving an individual of their entire potential future happiness, rather than merely the objects of happiness as a wife, children, income, or a home, which by definition are transient entities. Boethius, author of the early medieval *Consolation of Philosophy*, had lost all after his accusation of treason. However, he points out that the greatest loss is that of the opportunity to seek happiness, which is imposed by the extreme loss of freedom of his incarceration. It is this which constitutes the core punishment of jail: it deprives the individual of the opportunity to seek happiness in the world. In a prison cell there is no freedom of action, no freedom of association, but only a limited set of allowed behaviors. For all human beings, isolation one of worst forms of punishment, as seen in the case of the Puerto Rico political prisoner Oscar López Rivera.

The modern legal presumption is a rather presumptuous one, implying the right and power over all other creatures of the world—the rule of might rather than right. This clearly stands against the Christian view of man as 'steward' of the Earth. He is its caretaker, akin to a garden, who looks out for the best interest—an inherently altruistic definition.[86] The very same issues stood at the core of the debate between Gifford Pinchot and John Muir with regard to the policy which should rule wild lands in the United States.[87]

One contention with the current legal landscape is that the goals of the research are never fundamentally placed into question. There can be some slight modification, but only within an unalterable plan; animals in the end are absolutely required for scientific discovery. It turns Aristotelian virtue of biology into vice in the modern era. Moderns have complete access and control of a creature. What is particularly striking is the astounding scale of modern research. According to the Hastings Center, some 25 million animals are experimented on each year; PETA argues that the figure stands closer to 100 million animals in total.

We might turn to the history of science and technology to suggest new pathways of biological research.

The objects of astronomy, by definition, are unreachable objects, which in turn forced an enormous amount of creativity which culminated in the Scientific Revolution. We might similarly ask what would happen if biological objects were

---

[86] This was the Medieval view, wherein nature made for man's pleasure and use.

[87] Theodore Roosevelt's legislation sought to preserve many of the unique wilderness areas of the United States, as Crater Lake, a collapsed volcano which contains some of the most pristine water on earth. Muir adhered to a preservationist stance, or the 'idealist' view that we should completely set aside lands with little to none human involvement. By contrast, Pinchot adopted the conservation view, or the 'practical' view of managed care. It implied deep human involvement in the management of a forest; while we would use its resources, we would replenish these as well. Both views influenced Roosevelt during the turn of the century, leading to the creation of the US national parks—which are visited by more European tourists than minorities living in the continental US.

completely prohibited from direct experimental research; would biology necessarily decline as a result? The immediate presumption is that it would, but in fact we really do not know. It goes without saying that it would certainly force greater methodological creativity on behalf of its scientists. Many historical examples suggest the possible net benefit of such a policy.

As Johannes Kepler noted, it is usually not the rich who are creative, as their degree of comfort does not create pressure for innovation. His experience with Tycho Brahe more than aptly proves his point. We might similarly note that one critique of national economic development policies of states with an abundance of natural resources, which are relatively easy to withdraw and generate a revenue flow. This abundance of resources typically does not force the region to become efficient and productive by exploring the production of value-added commodities. Another example might be drawn from the world of telecommunications. Fiber optics was not invented by ATT precisely because it was so wealthy. The corporation could throw away millions of dollars on microwave tunneling, an inefficient and costly technology—which was problematic even when operating under controlled laboratory conditions. Since ATT had the money to burn, it was not pressured to this innovation, which was actually undertaken by Corningware, historically known for its kitchen glassware. Under financial pressure, Corningware ushered a revolution in telecommunications by creating a system whose bits could travels at the speed of light.

There are other examples whose objects are unreachable, and have thus been stimulated to unprecedented heights of creativity and scientific ingenuity. The core of Earth in geology is wholly unknown in the sense that it can never be reached; in spite of this seemingly preponderant obstacle, the science of seismology still emerged. Unique tools were developed to explore the earth's depths, particularly in Japan so characterized by the high number of earthquakes. While experimentalism constitutes a key aspect of scientific practice, is not the only criteria or method. Other research tactics exist, which by definition are not experimentalist. For example, the language of science is important, as can be seen in chemistry. For Lavoisier, who certainly performed many experiments, if language reflected nature, when correctly established it would reveal many of inner secrets. The result is our now well known chemical nomenclature and formulas which accurately describe reactions that need not be performed to know their outcome.

In light of these prior examples, we might ask whether biologists have had too easy an access to animals. Could it be said that biology is gifted with too many resources? Unfortunately, the answer appears to be affirmative. It is historically true that this access has led to large numbers of unnecessary deaths, which we will further explore in the book. We might suggest that it has taken too long to place substantive limits on biological research.

If we are truly going to talk about animals, lets talk about animals for their own interest—including as well animals for consumption. Our main focus in the next chapter is to answer whether the tide moving in a positive or negative direction . As Herbert Spencer, we must take the long view point of view, which does not contextualize issues in the time frame of a day, a week, or a month, but rather over a longer time span of decades (mesoscale) and even centuries. Is the tide of public animal morality moving in a positive direction or moving away from it? As Elihu Root pointed, out, it is not the minor play of the waves that counts, but rather the overall tide which is essential.

Matt Ridley provides some clues to these issues. Human cognition, as so much else, has been profoundly affected by evolution.

Experimentation and Animals in Science

Pliny (2379 AD), father of natural history, had been a solider in the Roman army under Titus, emperor Vespian's son. The two became close friends, which greatly aided his work in natural history, facilitating the obtainment of books and animals for his collections. Pliny noted that much of Rome's wealth, some 100 million sesteres, was wastefully used to obtain spices, adornments, and silk from the Far East. He lamented the many sacrifices he had made as solider. "Did we conquer the world for this?" He believed that the decadence of Roman empire was directly due to its success. While the important *Pax Romana* had established an enormous global market surrounding the Mediterranean Sea, its ultimate purpose had become a trivial and superficial one. Certainly there was more to life. Was purpose of humanity to be fat and superficially entertained? Such a shallow definition of progress would not do, noted Pliny the elder.

We might ask similar questions with regard to the role of experimentation in science and biology; to what ends and what purposes will biological experiments serve? Will biology be used for trivial ends, as a better perfume or fragrance or, as Aristotle noted, to understand the beauty of nature. Aristotle argued that reason was man's highest experience and his defining trait. One was happiest in the discovery of the underlying order of nature, and it is this trait which distinguished men from all the other creatures of the world. While plants reproduced, and animals were sensible and moved, men had the power of reasoning and understanding.

In the evaluation of experimentation in the history of biology, we have to make certain distinctions. There are many different types of experiments, not all of them have to do with animals.

While killing animals has been an important aspect of biology, the questions a biologist seeks to answer dictates the experiment undertaken, and in this sense blind routine experimentation is a waste of time. A whole host of experiments have occurred in biology that are not strictly 'animal experimentation' properly speaking.

One might question what value does bioethics holds for science, as some college students are wont to do. After all, how can it possibly contribute to the practices of science; does it not hinder rather than assist scientific inquiry?

Implicit in this characterization is the presumption that science operates outside ethics. In it, ethics is seen as a series of philosophical principles or ideas which may or may not be chosen and believed by the scientist. What the scientist publicly claims might be very different from how actually he behaves. Ethics under this characterization is merely a set of ideas 'in a galaxy far away', as in the movie Star Wars, no different from the mystical forces alluded to in it. The claims of philosophy, like Obiwan Kenobi's sermons, are seen as abstract notions without any real meaning in the world. In such a mind, no relationship between the two academic universes exist; both science and philosophy are discrete disciplines far apart from each other.

The relationship between ethics and science, however, is much closer than is generally presumed.

Science can be defined as a distinctly ethical outlook of the world, which is what makes it so unique. It can be defined as the 'search for truth' or the search for the hidden order of nature that is not immediately apparent to the naked eye; it is the quest to bring order where there is chaos. It is perhaps due to its distinct ethical outlook that scientists as a whole are often judged from a higher standard of behavior—specifically with regard to their work. The personal lives of scientists can be seen with much bemusement.[88]

---

[88] There are characters on one end of the social scale, as Richard Feynman liked to play jokes on his colleagues, to Isaac Newton who was distinctly antisocial. In this sense, scientists are very different from medieval monks in the expectations with regard to their public behavioral. However, when it comes to their actual work, the highest ethical ethics are presumed to exist.

Perhaps the key operating social presumption is that scientists as a whole do not lie, in principle. They might have strong disagreements, differ in their operating presumptions, but they do not purposefully deceive their colleagues. The sole aim of debate in science is to clarify ideas or to remove possible self deception. It is unquestionable that the notion of deception is patently hostile within the field, and the purposeful use of deception is one of the fastest ways to commit professional *hari-kari* (suicide). A few counterexamples provide a good illustration of these points.

There are many infamous cases where scientists grossly lied to the world, thus ruining their reputations. Hwang Woosuk, an alleged leader stem cell research at the turn of the millennium, claimed that he had created embryonic stem cells via cloning. Between 2004 and 2005, he was charged by the Korean authorities with embezzlement and fraud. By 2006, he had been fired from Seoul National University, and all government funding he had received was removed. Although a four year prison sentence was solicited by prosecutors, Woosuk fortunately escaped the harshest sanctions. His case is ultimately a tragic one, going from the uppermost top of field to its very bottom in a short span of time.

This is comparable to the cold fusion experiments of the 1980s. Martin Fleishman and Stanley Pons appeared to have discovered the golden chalice of energy in their experiments: free energy at room temperature. Although not necessarily cheating, it was certainly the case that they had not been rigorous enough in their analysis. Fleishman, who died in 2012, came to regret whole thing. In spite of the fact he had already retired by 1983, the experience ruined his scientific reputation. There is perhaps a similarity in kind, if not degree, to the creation of falsified data by Chinese scientists, who are relatively new to world stage. Too many Asian scientists have failed to internalize the ethos of science, and have thus severely undermined their reputation at a global level.

It is for these reasons that the ethics towards animals is particularly acute. Because scientists are judged by a higher standard of conduct and behavior than the rest of the population, egregious violations have a much greater social weight and consequences.

There is even a case to be made that science is tied to democracy. Its foundations began in Greece, as so much that is Western. The character of Greek society was ferociously democratic. All shared in the decision making process, from the "janitor" to the "Bill Gates " of their society. It is this dynamic which made it so unique, historically speaking. The similarities between the two mindsets of science and democracy can be observed in a comparison between Thales and Solon, who were contemporaries. Solon was a poet-politician who argued that men should not be enslaved for their debts—a practice that was common in that time period. Thales was a 'scientist-philosopher, who argued that all laws should be public, and easily accessible to anyone—something we take for granted today.

Both science and democracy have historically shared many similar traits. The Greek political process was based on reasoned debate: an analysis of positions, presumptions, and contradictions. Aristotle undertook histories which did the same; he surveyed all positions taken, had an internal debate with prior authors so as to clarify his own views and arrive at a conclusion.[89] Consider as well some counter examples. Tyrannies by definition are the 'rule of force' or of 'might makes right'. There can be no genuine debate under its political system, as the tyrant imposes his opinion by force. The manner in which science operates is clearly opposite to that of tyrannies. Science is driven and pushed forward via the free exchange of ideas, new data, debate, and new lines of research—notwithstanding its exceptions as in the case of Lysenko. Historically, consensus in science is reached by debate, and truth determined on its own basis as opposed to a political one. The procedure of science is problematic and contentious in that nobody has absolute power and nobody can impose by will of force. Truth is the outcome of the 'hive' of scientists in a particular field.

---

[89] It was in this manner which he helped to preserve Greek intellectual legacy for posterity, AND might be said first historian of science.

We may also point to the Greek notion of hubris, or excessive pride and arrogance. The goddess of arrogance was Hybris, whose husband was Polemos, the god of war. Polemos followed her wherever she went. In other words, reckless arrogance leads to the undoing of nations, much like Donald Trump today is doing to the United States. Note that this was also a common theme in Aesop's fables, and a core reflection of deep Greek values: humility before truth. This attitude occurs at individual level, but also at the social level as well.[90]

The relationship between science and ethics is much closer than it might appear to be at first glance.

Questions regarding the role of animal experimentation in biology appeared in is very beginning. A substantial portion of the history of biology has sought to answer the 'uniqueness of life': are living processes entirely distinct from nonliving or are they wholly separate? What accounts for the difference between animals and rocks? What is difference between the man that was alive 15 minutes ago, to the cold corpse that remains upon death? Although still the same individual, the constitution of the body is altogether different: its inertness, a lack of activity and sensibility, as well as the absence of heat.

Two principal positions have existed, that of vitalism, or the : notion that life was wholly distinct, and mechanism, or the idea that the inorganic world could be used to explain all living entities. For mechanism, there was no distinction between life and non-life, the former was simply a more complicated machine. Each view, debated until the nineteenth century, had their own distinctive traits.

For vitalism, the notion that living processes are inherently different to nonliving one typically required some external agent or force to imbue dead matter with life. All source of life is thus exogenous, even if there was a great deal of variation in the many different causes attributed to it. These causes could be

---

[90] These values might be contrasted to those of the nineteenth century. Common notions of the period included the belief in unlimited progress, and an utopian optimistic about future of humanity. There were a couple of problems in the formation of the industrial era that placed these presumptions into question. James Watt did not foresee the impact of carbon dioxide.

immaterial—God or the soul—or they could be material as well. In spite of its multiple shapes and forms, life was basically an imposition of order by an external agent, and hence the focus of most vitalists was the demarcation between life and non-life.

By contrast, in mechanism there no difference between living and nonliving process, and hence obtains a different character. While it might be hard to identify, there was a consistency of experience throughout the universe in the mechanist worldview, as in the case of fermentation which could be both chemical and physical. Fermentation was the same whether it occurred within or outside of it. The animal body could also be presented as a complex machine, full of levers and pulleys, pumps, pistons, and pipes. Mechanism's weakness was that it tended to oversimplify towards a crude reductionism. The aim was to reduce the complexity of life to a few simple key components that cut across all living organisms. Biology was ultimately reducible to physics and chemistry, a philosophy common in Germany. Vitalism focused on differences, and mechanism focused on similarities.

While we hold a mechanistic view today, should not presume that it consistently rendered the correct biological interpretation. The mechanistic view historically proposed many mistaken positions, as shown in the spontaneous generation controversy. It had been observed that maggots emerged from flesh, mosquitoes from swamps, and flies from rotting food as a banana or banana peel. Gonzalo Fernandez de Oviedo claimed that trees gave birth to animals. Mechanists believed spontaneous generation provided definitive proof that no difference between the organic and inorganic realms, a view influenced by Gottfried Wilhelm Leibniz. Known as father of calculus, Leibniz proclaimed 'monads' to be the source of life, and it is curious note the many similarities to his mathematical innovations (calculus). For him, the universe was made up of 'infinitudes within infinitudes' or monads, which were conceived of as atomic-like structures. Through this mechanism, Leibniz was able to posit an acceptable interpretation of life, given the limitations of visual technology at the time.

Leibniz was himself influenced by work of microscopists as Anton Van Leeuwenhoek, who constructed some of the best microscopes during the seventeenth century. His tiny well polished beads or 'microscopes' were very sought after, and with it he noticed particles moving in water which he referred to as 'animalcules' (protozoa).[91]

This body of work influenced Comte de Buffon who came up with the notion of organic molecules as being the source of life, which itself would influence Denis Diderot. Diderot believed his epoch was on the cusp of a scientific revolution in biology, fully aware of its consequential social implications. *"We are on the verge of a great revolution in the sciences... Do you see this egg? With this you can overthrow all the schools of theology, all the churches of the world"*. Obviously, this particular revolution did not happen. Leibniz's approach, proposed in the religious struggles of Europe, was an attempt to reconcile two diverse worldviews; it was the attempted towards the mechanization of the spirit, or an effort to 'scientize' Catholicism.[92]

Inversely, we should not discredit vitalism so casually as it has been a position taken by many important scientists and used to produce substantive contributions in biology. It has been pointed out that Aristotle was a vitalist. He believed that a soul animated the living and also made the distinction of living from non living processes. William Harvey, the discoverer of circulation, was also a vitalist in his belief that the process of life was unique—a notion also shared by Louis Pasteur. Vitalists in fact proposed many advanced scientific positions, and their refutation of spontaneous generation today constitutes a hallmark in the history of science.

Francesco Redi in 1668 undertook his important meat jar experiment. He noticed that when the meat was covered in a jar, no maggots appeared; if uncovered, maggots soon emerged. Furthermore, if he used a gauze, maggot formation would also be inhibited. In 1748, John Needham improperly boiled beef broth, and upon seeing microorganisms, used it as a proof of the mechanical interpretation of life, incorrectly believing he had had killed all organisms. However, when Lazarro Spallanzi repeated

---

[91] It is believed that saw bacteria, but did not recognize it as such.
[92] Leibnitz's notions were disproved by Robert Brown in 1852.

the same experiment in 1768, boiling beef broth for two hours, he succeeded in killing microorganisms. The processes of life were distinct from non-living phenomenon.

It would be ideal to suggest that the vitalism-mechanism debate concerning the origins of life was resolved only via experimentation. However, as the prior cases suggest, the issue was still being debated in the nineteenth century, and by then key differences between inorganic entities and organic life had been identified. Organics are characterized by irreversible processes. While inorganic metals and salts could be melted or dissolved, and then returned to their prior physical states, organic compounds could not be brought back to original state. One cannot uncook an egg, for example.

Yet there was an odd unity pervading all life. Organics are made up of three key compounds, including carbohydrates (high oxygen, soluble), lipids or fats (less oxygen, non-soluble), and proteins (amino acids using nitrogen).[93] Because all animals are made up of the same compounds, each in turn can serve as nutritious food for the others.

There was an even deeper complementarities between animals and plants. Stephen Hales, author of *Vegetable Staticks* and inventor of the pneumatic trough, noticed something particular. While animals breathed oxygen and expired carbon dioxide, plants did the exact opposite, 'breathing' carbon dioxide and 'exhaling' oxygen. Each group of living entities provided what the other lacked, again suggesting a unity to life that had not been previously detected. None of these studies had been undertaken with animal experimentation.

But the debates continued, as the development of chemistry showed. In their experiments using a calorimeter, Antoine Lavoisier and Pierre Simone Laplace showed a parallelism between living and non living phenomena. Combustion in fire was similar to the process of respiration in animal. both used up oxygen and had carbon dioxide as a byproduct, suggesting an equivalence between nonlife/life and a potential proof of the

---

[93] Note that the term 'amino' refers to material derived from ammonia (nitrogen).

mechanical interpretation of life. It is important to point out, again, that although animals were used, it was in the mutual interest of both experimenter and experimented that the animal survived. No harm ever came to the animals used in these particular experiments due to the character of the experiment.

The mechanical view continued to gain ground during the nineteenth century, as when Frederic Wholer synthesized urea by heating ammonium cyanide in 1828. Urea, a component in urine, was not strictly organic. Similarly, Claude Bertholet synthesized many other compounds, which had been previously believed to have been only organic: methane, benzene, methyl alcohol, ethyl alcohol, among others. It became increasingly clear that chemistry was an intrinsic component of biological phenomenal. It is again important to point out that these experiments, so critical to the understanding of life, did not require animal experimentation.

This was not to say that animal experimentation played no role. Food was much more complex than previously believed. Japanese soldiers suffered from 'beriberi' because they only ate polished rice. Christian Eijkman studied the issue in Java in 1886 in a chicken feed experiment, feeding certain groups whole husks and others only polished rice. The latter developed beriberi. However, so powerfully influenced was Eijkman by the ideas of the time, specifically the germ theory of disease, that he wrongly interpreted his result due to an antitoxin in the food, which he believed was eliminated during its processing. Johns Hopkins and Casimir Funk disagreed, and develop the vitamin hypothesis in 1912. When dogs were fed only sugar and starches, they died. When rats were only given synthetics of carbohydrates, fats and proteins, they also died. However, upon the administration of milk, these recuperated. Somehow, milk had an extra ingredient necessary for life, which they ended up calling called 'vitamin'.[94]

Without a doubt, animal experimentation and the consequent fatalities have contributed to knowledge in biology, yet we may ask whether there is a cost involved. A whole host of views have existed with regard to the role of animals in biology. Aristotle

---

[94] Vernon McCollum introduces the 'letter vitamins', vitamin E being the first.

believed that the study of dead animals was not appropriate for biology; the study of living processes required the study of living creatures as dead ones would only yield information on death. He did point out the difficulty of observing dead bodies, in that their lack of blood flow could be deceptive.[95]

Medieval views generally opposed animal vivisection. Thomas Aquinas argued that cruelty to animals would dull the heart and foment cruelty to man. If a child hit a dog, they would end up becoming cruel adults. St. Francis, who took vows of poverty, extended the notion of human rights to animals; one had to treat animals with dignity and kindness.

As science advanced, there can be no doubt that animals have played a role in biological knowledge, in particular the cruel and torturous practice of animal vivisection. But its role here is somewhat problematic. William Harvey discovered circulation of blood using this technique. He would place rabbit on a board with paws firmly tied, opened its chest cavity, and modified its internal organs, such as the tying of the aorta to see how it worked. As the King's physician, Harvey had ample access to many of the animal specimens the king liked to hunt, including dear and bears. The countless number of such specimens does raise the question of how particularly useful these were. The proof of circulation however came not from the work of vivisection but through that of quantification: the measurement of blood.[96]

It is perhaps significant to note that Harvey did not have a good reputation in life, and one has to wonder the degree to which animal experimentation might have played a role in the creation of his public image. His claim of blood circulation was not initially believed, even by individuals willing to do so as

---

[95] Aristotle himself was fooled, concluding that the brain contained no blood, and hence served only to cool the body. For him, the seat of the soul was the heart because it heated the blood upon its expansion (referred to as the diastole). Today we know that motion of the blood occurs on the systole (contraction) of the heart.

[96] Harvey estimated that the amount of blood which flowed. If the left ventricle holds two ounces and pumps 72 beats per minute, this would equal 540 lbs of blood—an amount impossible to be created and produced by the body.

Gassendie. It was not until Gassendie had mechanical proof in Torricelli that he finally then came to believe the claim as a credible one. The ultimate proof, however, was provided by the microscopists, particularly Marcello Malpighi. In his study of frogs with the microscope, he was able to identify capillary structures connecting veins and arteries. Malpighi himself claimed that he had sacrificed an entire race of frogs in his experiments.

A great many animal vivisections were also conducted at the Royal Society. Robert Hooke and Robert Boyle tied dogs, removed their thorax and chest wall so as to be able to observe the lung and heart in action. They used bellows to pushed air into the lungs, showing that the heart kept beating and that the animal surprisingly remained alive. Hooke himself detested this sort of experiment, as it was fairly obvious that the subjected animals were suffering a great deal in the days prior to anesthetics. When Richard Lower contacted Hooke to repeat the experiment, Hooke initially refused. Lowe wanted to answer whether it had been the air or the action on the lungs which kept the animals alive. It took Lower three years before Hooke finally relented. This time around, they used a two-billow apparatus, which kept lungs continually full of air. It then became clear that it was not the action of lung but the air per se that kept animal alive. The dog's heart continued to beat . . . but obviously not for very long.

Hooke's mentor, Robert Boyle, was a key leader in experimentalism, and is today known for his pressure law.[97] He established key epistemological principles, such as the notion that theory must emerge from experimental data and cannot proceed independently of these. He certainly was an experimentalist rather than a theoretician, and never came up with a comprehensive scientific theory. In those days, ethics, epistemology and science were all considered mutual branches of philosophy.

---

[97] He worked a great deal with the air pump and the creation of vacuum, continually testing effects with all sorts of creatures. Once he placed a viper, and noted that upon evacuation of chamber, it became quite 'agitated' before 'becoming extinguished'.

Yet his views on animal experimentation were somewhat mixed. He recognized that working on humans was unethical and also that animals did suffer as well. In spite of this, he continued to sacrifice a great deal of them in his vacuum chambers. He appears to have rationalized away true character of animal experimentation. In his essay on the ethics of animal experimentation, he often sought to relieve the animal's pain, but did not consistently do so in practice. Once, while experimenting on a bird which was obviously distressed and nearing exhaustion, a horrified female observer stopped the experiment shy of the animal's death. One may here point to a substantive change when the locale of the experiment is shifted from the private to the public realm. The public, as well as the scientists involved, became more aware of the actual moral nature of the experiment when these are performed in public.

The opinion of others often acts as our own moral referents, as depicted in the painting by Joseph Wright of Derby. It is a realistic portrayal of actual witness reactions to Boyle's vacuum experiments. It is an enormous painting, placing the viewer as if they were in same room with the experiment. That Wright was an asthmatic may account for his sympathy with the depicted. In it we see a darkened room with a vacuum tube with a bird in the middle, and a scientist surrounded by a small audience. Two young lovers are totally oblivious to the experiment. The young male kids are curious and entranced by it, apparently gaining a sense of identity and awe about the natural world. By contrast, the girls in the room are decidedly affected by the suffering cockatoo empathizing with its loss of agency; however much it might try, the bird could not escape the fatal vacuum.[98] The older man is contemplative, likely given that he is closer to death as the bird than the others. Boyle's work was mocked by Jonathan Swift in *Gulliver's Travels* (1726), where scientists undertake seemingly endless, cruel and purposeless experiments.

Natural philosophers during the seventeenth century actually showed a great variety of opinions with regard to animal experimentation. The natural historian John Ray believed that all

---

[98] Cockatoos are known to well imitate the human voice.

animal experimentation was immoral. For him, animals were God's creatures, which revealed His goodness and existence. As such, vivisection an act of blasphemy. On the other hand, Vesalius, known for *Re De Frabrica*, actually believed that vivisections were morally beneficial, a somewhat sadistic claim. He would dissect the pregnant female of a species, only to note how the mother would tenderly protect its unborn fetus by seeking to ease pain over their own anguish. Stephen Hales also performed a great amount of experimentation with dogs (college) and horses, most of which were undertaken in his own home. The cries of animal pain heard by his neighbors not only aroused suspicion and rejection but led to his communal shunning. The writer Alexander Pope, his neighbor, specifically mocks him in the poem 'Against Barbarity to Animals'.

Albrecht von Haller, a noted vitalist who came up with notion of irritability and sensibility as proof of vitalism, tortured more than 200 animals in order to prove his theory. These were exposed to acid, fire, and noxious substances to observe their reactions. His exploration of the nervous system was particularly cruel, and we might ask how scientifically useful his experiments were. Germany had a long reputation for reckless cruelty in biology prior to WWII and the routine use of prisoners for experimentation.[99]

In conclusion, animal vivisection, the most cruel treatment to another species, has played a part in the history of biology. Many important biological features were revealed in the process. Boyle recognized that it was far better to experiment on animals than on human beings. Yet, we might raise the question with regard to the degree to which vivisection actually contributed to scientific advancement in biology. The various cases reviewed suggest that it represented great deal of wasted effort. Harvey's vast number of vivisections did not actually prove circulation; rather, quantification did. Boyle's testing of his air pump killed thousands of animals; yet it might be asked how many animals

---

[99] Incidentally, Claude Bernard specifically rejected use of anesthesia when studying the nervous system, claiming that it might in some way alter its results.

need to die before the discovery that all used oxygen? Hales' use of horses for blood pressure proved ruinous for his local reputation.

It is important to note that most of these experiments were performed in private rather than in public arenas. Public vivisection would have undoubtedly exposed the cruel treatment to which innocent creatures were being exposed, and certainly have increased the public's moral acuity towards experimentalism. As in Derby's masterpiece, the public would have shown the same reaction as that shown by the young women. Our gut reactions reveal that our minds are not blank slates but rather have cognitive biases built within them.

Vivisection, or more broadly animal experimentation, is only one of a multitude of aspects of scientific innovation during the period. Biology did not advance solely from animal dissection, but other methodologies were equally significant, as the weighing and measuring by Harvey. The synthetic creation of organic compounds also helped clarify many issues. We should also not overlook the obvious: the simple but important role of debate and theory. What you are going to observe requires theory, and in this the process of debate itself was critical to advancement; this is what principally characterized pre-Socratic philosophy.

It also clear that we no longer have to repeat the same experiments, and hence point to the fact that there exists a cost to the acquisition of biological knowledge. Biological forms are so complex, that they have to be 'broken apart' in order to begin to identify their underlying structures, and in the process countless animals were sacrificed. With regard to biological systems, life arises from complexity; once destroyed, it cannot be put back together. This fact helps explain why history of science is so important. Akin to the teachings of the Hippocratic treatise, we can directly learn form the collected experiences of other individuals; we do not need to relive everything first hand to acquire such knowledge.

Part II: The Moral Animal

Introduction to Sociobiology

We might begin by first asking why one should study sociobiology in an ethics class? To answer this question, it is useful to first look at the history of lenses.

Newton's groundbreaking study of light used the process of analysis and synthesis to discover that light is made up of colors. This discovery, in turn, led to groundbreaking improvements in their design, and is still the principal model used to this day. Reflective telescopes as these eliminated chromatic aberration, the fundamental flaw of all prior refractive scopes. Lenses are our window to the starry universe, and their imperfections produce a distorted view of the cosmos.

Similarly, human eyes are the window to the soul in more ways than one can imagine. In his study of eyesight and light diffraction during the medieval period, the Islamic scholar Al-Hazen realized that to understand sight, one need to include the brain. Damaged brains produced distorted images of the world, not unlike the distortion of gravitational lensing produced by massive stellar objects, first described by Einstein.[100] Once we are aware of this distortion, astronomers can create 'corrective telescopes' to compensate for the error.

The prior examples suggest that a scientific understanding of cognition will allow us to detect previously imperceptible mental distortions, as the chromatic aberration which once afflicted all early telescopes. While we now take it for granted that distorted corneas can be quickly remedied by corrective eyeglasses, we tend to give little attention to our mental lens. Just as a nearsighted person is unaware of how blind they are until they put on eyeglasses, unless we critically analyze our own human cognitive structures, we will remain ignorant of our hidden biases.

---

[100] Light bends around heavy objects, hence creating distortions in astronomy.

We might ask, for example, whether 'taste' is universal. Are all things that taste good to us, taste good to other animals as well? Is taste, as color, an inherent function of the object eaten or of a subjective relation to eater? In fact, we tend to prefer things which are beneficial; beneficial foods 'taste good' to us. A good example of this phenomenon are spices, most of which have some sort of antibacterial function: rosemary, oregano, salt, and so forth. It would make sense that there would exist a close relationship between taste and benefit; things that taste good would be more prone to be eaten and hence encourage the health and 'genetic fitness' of its consumer. By contrast, rotting meat or excrement is bad for us due to its high bacterial content, and hence we tend to be repulsed by these. Their cognitively perceived bad taste is beneficial to us because their avoidance increases our longevity. Note, however, that rotting meat and excrement taste good for flies, which they quickly detect and are immediately attracted to it. For them, rotting meat is a tasty snack for similar genetically beneficial reasons to the fly. Inversely, for the worm that grows and lives in meat, salt does not taste good at all as it alters body osmosis.

In other words, taste is universal only to the particular species at hand. It goes without saying that taste is a subset of cognition.

There is no doubt that cognition is not universal across species. Any dog owner will readily observe that dogs do not take the same sensory inputs as their owners. Information is not merited the same value by dogs as humans. A dog will be interested in a fire hydrant for different reasons than its owner, as its functionality of putting out a fire. Dogs focus on certain objects and not others, a phenomenon referred to as cognitive bias. For example, dogs will tend to intensely seek other dogs. They do not smell leaves for sake of leaves rather but to smell markers left by other dogs, given that urine is each dog's 'presentation card'. Another well known canine behavior is the tendency to quickly jump on the grossest smelling things. On a surf trip to "*La Selva*" in Fajardo (Puerto Rico) my mutt one time jumped onto a mound of cow feces, as if it were jumping into a swimming pool. While repulsive from our point of view, this behavior makes complete Darwinian sense: to hide its own smell, thus making it much easier to sneak up on a prey. Dogs have a 'cognitive bias', which lends itself to a particular behavior which is ultimately evolutionarily beneficial to that species. As a consequence of its improved hunting ability, dogs were able to successfully spread their genes to future generations, also passing along successful traits.

The key of evolution is 'fitness'; if biological entities do not pass genes on to future generations, then the biological form 'ends' (becomes extinct). As Aristotle noted, individuals are impermanent; we all die at some point, unfortunately. What is permanent in biology, however, is the species, which for him was universal; roughly similar forms propagate over the long eons of time. Aristotle is obviously no longer alive today, and hence why reproduction is so important. If one leave no descendants, then the inter-generational line is cut. Much of history, however, is in fact a record of severed generational lines; it is the tale of shapes and forms that did not make it for some reason or other. Most lines of descent do not make it, either for exogenous or endogenous reasons.

Returning back to our original question, if we agree that the human brain is the 'cognitive lens' to the world, it then follows that the study of the 'evolutionary mind' is essential to understanding decision making and ethical judgments regarding to what is 'right' and 'wrong'. It would provide us with a good understanding of our particular cognitive biases, as well perhaps to lead to realistic solutions—a lack of which usually leads to poor policies.

We can take the Catholic Church as an example. All Catholic priests are symbolically 'married' to Jesus Christ. This aspect is a laudable goal in that it recognizes the ruinous role sex has played in human interaction.[101] As a result, all priests, in principle, are celibate. Unfortunately, as is routinely revealed, sex is a key component of the human psyche, which helps account for the high incidence of pedophilia in the Catholic Church.[102] The

---

[101] Men all too easily abandon their responsibilities because of it—one of early adulthood's first costs.

[102] The most recent incident (August 2018) were the pedophilia charges of some 300 priests throughout Pennsylvania.

human sexual urge will always exist; socio-legal sanctions against its natural manifestation only forces its illicit display. In the long run, the Catholic policy is a foolish one as it goes against a fundamental strand of human nature, regardless of whether we agree with its Freudian interpretation or not. The importance of reproduction is genetically universal across all living entities. Islam has its own prescriptive variant abnormality, similarly established for beneficially purposeful social ends.[103]

To obtain the most realistic assessment of human behavior, we must first turn to sociobiology, also known by other names as 'evolutionary psychology'. Just as the telescope is used to observe light which was generated millions of years into the past, the use of sociobiology allows us to look back into the cognitive legacies which were developed eons ago and profoundly shaped the human condition. The deeper into the past we look, the hazier our image becomes, but the more profound and substantive our understanding becomes.[104]

The behavioral impact of human evolution is a difficult topic, as suggested in the introduction. There are an abundant number of oversimplifications, which entirely ignore its insightful points and interpretations. Similarly, the politics of policy such as the rise of feminism have also injured its public reception; authors as E. O. Wilson were wrongly accused of misogyny. Human behavior is the complex outcome of many factors. This section will not pretend to identify all human cognitive biases, but rather discuss some of sociobiology's core ideas. Again, it has to be emphasized that the topic is a deceptively simple one, touching the realms of psychology and psychiatry. However, biologists

---

[103] Recognizing that sex affects the interaction of men, Islam keeps women hidden from view within their *burca*. While there are many variations of the form, some covering only the head and others the entire body, all basically reduce the temptation of going against marriage vows. To a degree, the *burca* is an implicit contract between males, which helps exchanges to be less aggressive. In the system, males are less likely to perceive others as being in direct competition, which leads to smoother social interaction by reducing direct sexual competition.

[104] The best overview of sociobiology is that found by Robert Wright, *The Moral Animal* (1994).

have provided new interpretations, which include new and unique factors which historically have been ignored by the two disciplines.

It goes without saying that the field is an extension of the biological paradigm; revolutions as such are usually defined in retrospect.[105] There is no doubt that a revolution in social sciences has been silently occurring. The Darwinian paradigm is being extensively applied across many fields in the humanities and social sciences, as sociology, economics, and politics, among others; it is a work that is still relatively new and in its early phases. It is truly a fascinating period in intellectual history where the cross pollination of ideas from different fields is yielding fascinating results. The way one looks at humanity will be forever changed.

Let us now turn its history, starting where we left off in the preceding chapters.

As noted before, Darwin was unable to account for two things in his evolutionary work: the nature of heredity and altruism. This gap in his theory explains the long delay in the publication of his final work, the basis of which was written more than a decade (1844) prior to its final release (1859).[106] He recognized its social and philosophical importance, and had given a copy to his wife Emma, instructing her that she should publish the work in case of his unforeseen death. He had reasons for concern, as Darwin had returned a sickly and frail man from South America. As Spencer, it was impressive that he accomplished as much as he actually did.

Darwin's mother died when he was but 8 years old, and his elder sisters played a big role in his rearing. At one point in his youth, he wanted to start an insect collection, to which his sisters opposed for its wanton injury to creatures. The young childhood experience might help account for Darwin's strong conscientiousness. Darwin fundamentally disagreed with social

---

[105] The Scientific Revolution of today was not fully appreciated, and only coined later in
1950 by Alexandre Koyre, Herbert Buttefield, and others.
[106] Mendel's work was seen as irrelevant to a degree, focusing on the explanation of stability rather than Darwinian change.

Darwinism, noting in that it could be used to justify all sort of injustices. The notion that society became merely a jungle 'survival of the fittest' was morally problematic for him, as he noted in a letter to Charles Lyell.

A key failure of his theory was Darwin's inability to correctly identify the scale at which evolution was occurring. He could only account for things as altruism, or selfless behavior, if he presumed the process occurred at the group level, defined as interspecies competition. At this scale, because the collectivity benefited from an individual's sacrifice, it became an important adaptive tool of group survival; successful future progeny went on to reproduce. However, the main problem with his interpretation was that it could not account for intraspecies competition. Why should a particular trait be passed and predominate in the first place at all if the individual did not directly benefit from it? As such, the trait would not be passed down between generations. There was no clear mechanism or explanation for this dynamic in Darwinian theory.

One surprising aspect of Darwinism is that Darwin himself at one point had been a 'Lamarckian', adopting the notion that each succeeding generation acquired the traits that had been gained by parents through their activity. In his early career, however, Darwin cmae to detest Lamarck's presumption regarding knowledge of its finality and ultimate goal: the idea that evolution tended towards greater complexity and perfection.[107] The Darwinian model was clearly against this, and rejected its teleological explanations. For Darwin there was no end and no purpose to evolution; creatures just adapted to their environment, and this aspect of his theory went against the religious dogma of his day—another reason for the delay. Going against large social actors was not something that should be trivially undertaken.

The attempt to apply Darwinism, or evolution via natural selection, to understand human behavior has been fraught with problems. A later effort was undertaken by Desmond Morris during the 1960s with his popular books as *The Naked Ape* (1967) or *The Human Zoo* (1969). Morris was drawn to the field

---

[107] Darwin's' grandfather, Erasmus, also held teleological presumptions.

by his early study of culture in chimpanzees. He had chimps draw and when he compared these to the drawings of human children, Morris noticed certain general consistencies. While chimpanzee 'art' looked like gibberish at first glance, these tended to balance the white space in a drawing. For example, if the paper had a box on the left portion of the page, the chimpanzees would draw on right side of the page, presenting a consistent symmetry. By contrast, while human children drew the same patterns, they showed quickly advanced and began drawing outline of objects, showing higher cognitive sophistication and a level of representative abstraction which which the chimpanzees never produced.

Morris's works have many interesting insights. For example, a creature's neurological complexity is profoundly influenced by its environment, or specifically its food source. He compared the behavior of a koalas and sloths to that of dogs. Creatures with stable food sources as the koala's eucalyptus trees, had slower paced neurological processes. However, creatures with unstable food sources tended to constantly seek sensorial input, which obviously aided in their survival. As dogs are omnivorous and do not know where their next meal will come from; it is to their benefit to constantly look for food: rotten carrion, opportunistic prey captures, and so forth. Constantly moving and seeking sensorial input greatly aided their survival as a result.

For Morris, humanity is a "naked ape" because, in contrast other creatures, we do not have specialized biological tools: no wings, claws, and so forth. The only tools man has available is his brain and hands, but it is these which give humanity so much freedom, for it implies he can adapt to many different circumstances and locations—unlike many other animals. For him, men also now lived in human zoos.

In a modern society, social groupings tend to be very large, and these sizes tended to lead to consequent increases in violence given the reduced opportunity of leadership. The main leader in a corporation is the CEO; secondary leaders have less power and influence. Morris's observation can be easily understood by considering the following example. A premodern region of 1,0000 persons might be composed of 10 tribes of 100 individuals each, thereby having 10 de facto leadership positions. The same number of individuals might make up a corporation, with only 1 leadership position, which in turn implies that nine males would be left without significant social power and influence. For Morris, such social settings were pathological, going against human nature, and hence the increase of crime in modern urban landscapes. As the positions of power males could occupy had drastically declined, the outcome was expressed in what today we regard as criminal behavior.

There can be no questioning of Morris's central observation: that human evolution occurred in a very difference context than it does today, thus leading to consequent maladapatations. All men seek power, but few have opportunities to gain it in a modern setting.

In spite of its insightfulness, one of the critical problems with Morris's ideas is that they too were also teleological. Evolution never knows what future conditions will be, and thus is not a predictor of future forms or future behaviors. Evolution only acts at the moment, within its immediate circumstances; and it is these circumscribed conditions which dictate its particular notions and criteria of success and growth. While Morris's ideas could have led to a potential revolution in biology, these were never accepted as such in academia due to their Lamarckian character. That being said, his books and documentaries were nonetheless extremely popular, and suggested that a new way of looking at human nature was to emerge just around the corner.

The rise of sociobiology represents the most successful attempt at providing evolutionary explanations of animal behavior—of which humanity is included. More rigorous than Darwin's early ideas, it avoided the determinist trap of social Darwinism and in fact revealed many of its contradictions. In contrast to eugenics, which contrary to its name has no basis on genetics per se, sociobiology's genetics foundation produced innovative theoretical advancements that were not foreseen at the beginning of their intellectual endeavors.

There were many who participated, but we will mention just a few key figures. George Williams's *Adaptation and Natural Selection* (1966) set the basic infrastructure of sociobiology and the new ground for future developments. The two young scientists who most advanced the early field were William Hamilton and Robert Trivers. Hamilton was a mathematician whose original statistical analysis led to groundbreaking ideas, some of whom were plagiarized by John Maynard Smith. While on an expedition to Brazil, Hamilton died at the age of 63. For his part, Trivers was then a student of E. O. Wilson, with whom he did a study on island biogeography: what would happen if an micro territory was wiped completely of life? Trivers laid many key contributions, and still remains active to this day.

While his mentor E. O. Wilson did not invent sociobiology, he certainly popularized it with a 1975 book under the same name.[108] Wilson, a US southerner, was incorrectly seen as attempting to revive social Darwinism under a new scientific guise, and at an meeting of the American Association for the Advancement of Science (AAAS), students poured a jug of water over his head while chanting "you are all washed up." Some the most ardent criticisms came from fellow Harvard scientists as Richard Lewontin and Steven Jay Gould, a paleontologists who produced some of the best popularization of natural history that exist to this day. Their concern was that sociobiology would end up becoming a new eugenics, used to justify social repression so typical of the German Nazi period. Some of these criticism were faulty, and presumed its authors were arguing that human nature was fixed, and hence that the institutional molding typical of the '*tabula rasa* school' would decline as a social policy. Was sociobiology merely reinforcing existing prejudices or was it providing truly new and novel interpretations of human nature? Some of Wilson's early comments which suggested the validity of his detractors, were soon retracted.

Richard Dawkins also helped to bring the field to the public with his synthetic book, *The Selfish Gene* (1976). Another popular book, it described some of the key results of Hamilton and Trivers' work, and might perhaps have been too far ahead of time. It was attacked by Philip Kitcher, who somberly noted that while astronomical theories have no social consequences,

---

[108] *Sociobiology* (1975)

biological ones of human nature do. Even if Dawkins did not turn the notions into an ideology, someone else likely would, noted Kitcher. If, perchance at some future date, the theory was then shown to be wrong, it would have had even more grievous consequences that the eugenics driven Nazi movement.

Does this mean that should not theorize at all about the impact of evolution on human nature? The best synthesis of the field, Robert Wright's *The Moral Animal* (1994), clearly showed that such concerns are misplaced.

George Williams clarified and extended many of Darwin's ideas. Given that evolution does not know where it is headed, we must then look at a creature's immediate context to understand the forces influencing their behavior. What are the immediate benefits of a particular change? Perhaps William's fundamental contribution was his shift of focus; the unit of evolution rested neither at the social (group) level or the individual, but rather the gene. It was here where the sociobiologist needed to turn to understand the biological-evolutionary underpinnings of behavior.

Upon this basis, William Hamilton came up with the key notions of *kin selection* and *inclusive fitness*, which in short means a preference in behavior for related individuals. Altruistic behavior was perhaps a misnomer in that 'greed' underlined all altruistic behaviors as they were a direct function of relatedness. For example, social insects often share ¾ genes, and are more akin to clones of one another than 'human siblings'. Animals showing the greatest altruism shared the highest commonality of genes. While two siblings sharing ½ of their genes will compete intensely for parental resources, they will also come to aid the other in a time of need.

Robert Trivers took Hamilton's work as his basis of analysis, applying these ideas to the interpretation of human families. Human behavior showed many similar traits to other animals. What Williams referred to as *sacrifice*, in common parlance *valour*, and Hamilton as *altruism*, Trivers redefined as parental investment—or specifically *male parental investment* (which we will call MPI).

The existence of a gender preference which varied greatly between classes was particularly odd, but whose reason becomes obvious when seen through the eyes of the genes (sociobiology). Boy and girls are given preferential treatment according to the social strata to which their families belong. Males tend to be favored in the upper classes, as greater wealth allows for a much broader extension of a family's gene pool *vis-à-vis* a female's limited reproductive potentiality. On the other hand, females tended to be favored in the lower classes, as these were seen a guaranteed genetic investment. A male belonging to a lower class has the cards stacked against him, and hence a lower probability of reproductive success.[109] A pretty female belonging to the lower classes tends to marry upwards with a member of a higher social strata, and hence a better 'genetic investment' for her parents.

While altruism had been a problem for Darwin and Morris, the new school of sociobiology provided a successful interpretative framework within the evolutionary paradigm. Only by showing how behavioral changes were in an individual's genetic best interest, could one then demonstrate why such a trait would then spread in a population. Darwinian allusions of blind 'group benefit' could not account for how such a trait would have spread throughout a population in the in first place. The genetic analysis of behavior, when seen from this lens, provided not only more rigorous explanations, but were also relatively good predictors of such behavior.

We have to make a few disclaimers. The first is that genes do not control behavior directly. As pointed out by Dawkins, whereas behavior has to be, by definition, quick and rapid,[110] the genetic coding for proteins is a slow chemical process. The second disclaimer is that genes do not miraculously predict the future. The long term ancestral environment during the last million of years is key to understanding the impact of human evolution, but it is not the only one. Here, the 'environment'

---

[109] Females are naturally opportunistic in the sense they typically seek relatively higher resource males, whom will tend to produce better opportunities for their progeny.

[110] A reaction to an attack by a lion for example.

actually consist of two distinct realms: the natural and the social contexts. Evolutionary driven behavior is essentially a compromise of demands between the two. Factors include the resources of the natural environment, or the typical natural selection, whereas the social environment, consists of the whole host of social relations surrounding the individual. Both natural and social environments will push and pull in slightly different directions, either serving to aid or undermine genetic success. The outcome of evolutionary driven behavior will thus lie in the interaction between genes and the environment—genes are non deterministic, allowing for a great deal of plasticity and flexibility in human behavior. While humans are not completely blank slates, neither is their character completely written in stone.

Finally, it is important to note that sociobiology does not deal exclusively with human behavior but is used for to account for behavior of species across the entire animal kingdom—and as such it is a powerful predictor of behavior, human or otherwise.

One useful example of its predictive value can be seen in accounting for the different behaviors of the sexes, each of which could regard the other as a species from a different planet. Each gender unknowingly tries to use one another as means to their own ends; both are mutually taking advantage and mutually deceiving the other. An anecdote by Calvin Coolidge is perhaps apropos. While visiting a chicken farm, his wife inquires about the number of times roosters mated per day—the answer to which was more than twelve times per day. Mrs. Coolidge then has her attendant make sure her husband receives this information. Upon hearing it, Coolidge replied, "with the same hen?" The answer was that roosters mated with different hens each time—to which President Coolidge noted, "Tell THAT to Mrs. Coolidge."

There are different reproductive strategies between the sexes. The rates of reproduction of each are vastly different. The *Guinness Book of World Records* shows that a man from Morocco had a total of 888 children, thus setting a new record for male fecundity.[111] By sharp contrast, because a woman can only become fertilized once a year, by the end of her reproductive life

---

[111] It has been claimed that Genghis Kahn had many more.

cycle at some 54 years of age, the maximum she could achieve is some 30 children at best. In other words, human males can reproduce at a rate approximately 33 times that of women—if not higher.

The costs of reproduction also vary greatly between the sexes, including the particular cost of their respective gametes. The female egg is physically enormous in size relative to the inseminating sperm. Onto this is fact we might include the burden of rearing and child care typically falls upon the female, an added cost from which it should be distinguished.

The enormous variation in the potential rates of reproduction result in variations of their reproductive strategies between the sexes. Females in particular will tend to be 'coy', or choosy when picking a mate—and will tend to focus on a male's resources. However, resources are not the only criteria used, in that 'wealth' is not necessarily defined by income per se. The criteria of fitness used depend on their operational context. Wealth in tropical lands will be measured differently from wealth in arctic climates, for example. It is somewhat striking, however, that even if a woman is wealthy or holds a highly remunerative job, she will still seek a 'high resource' male even when such concerns are irrelevant to her actual needs. In this case, the evolutionary legacy of the ancestral environment exists out of context with a woman's actual financial reality.

Men, on the other hand, will be 'eager' to initiate and push a relationship forward. The pace of a relationship will generally be too slow for the male, and a humorous anecdote illustrates this point. Male turkeys have been found trying to court and mate with a stick attached to a plastic head. Human males, however, are well aware that resource allocation plays a key criteria of female selectivity, and hence during their twenties, will tend to focus on the obtainment of status and resources. Darwin, incidentally, did just this. His trip to South America in the 1830s led to a considerable boost of his reputation, quickly rising in status and being more widely sought after. This higher status, by definition, rendered itself to a relatively wider range of available mates than if he had never set foot outside of England.

Humorous worldwide courtship rituals also illustrate these dynamics. Male animals show genetic fitness in the most bizarre of manners, such as the Australian bower bird's structures and ornithological performances involving acrobatic feats or a parrot's use of artificial plumage to attract mates. Human courtship rituals can be equally amusing, as the facial markings and odd gestures of the Woodaabe Tribe in Niger. Again, it is important to point out that "fitness" has many different meanings, relative to the species and environmental context; physical fitness is often seen as a marker of genetic fitness hidden underneath— and of the ability to pass genes to future generations.

There are some bad news for the ladies: the rate of monogamy varies greatly across species and across the natural world, in particular those of mammals. These tend to be very low, typically hovering around 4%. In these cases, males are typically much larger than female because their MPI very low. Males hence tend to follow 'eager' strategy, and contribute very little to child rearing. The rates of monogamy in primates are somewhat higher, at 18%, but still considered low in the animal kingdom. Generally speaking, males throughout the animal kingdom have a very low MPI, having little incentive to 'stick around' after mating.

MPI indicates the degree to which males invest in their progeny—the exceptions of which prove the rule. The few species with high MPI incidentally show behavior traits which are inverted from their usual counterparts, but still follows the logic of genetics. Males might carry eggs (sea horses) or take care of infants (penguins), for example. Because this activity is a costly one, males in such species tend to be the 'choosey' ones with coy behavior; inversely, the females of such species reflect typical male traits, as less selectivity in mating and greater coloring.

However, one cannot draw models and ideals of behavior from other species with very different genetic history or environmental contexts. Monogamy is in fact one of the most common worldwide human attributes, found in 94% of all relations. The only cases where it does not exist is are dictatorial societies, or in polygamous societies with a high degree of inequality between the sexes. Monogamy implies a level of courtship and voluntary decision making, which is also common to nearly all birds. Since birds cannot leave their eggs alone, as these will be quickly eaten by the nearest predator, both parents play a shared role to preserve their costly genetic investment. Were it otherwise, the species would die out. A similar dynamic occurs in human mating.

Human male parental investment is typically much higher than in other species for various reasons, the principal of which is bipedalism and the consequent large brain which evolutionary followed. After birth, primate young tend to be well formed and functional, able to jump on their mother's back. Human babies, on the other hand, are born premature and highly dependent on their parents; they are 'naked apes' wholly unable to defend themselves. Without MPI, human infants would quickly become 'tiger snacks', as the eggs of most bird species; for all sakes and purposes, babies are the functional equivalent of bird eggs.[112] MPI thus helps to increase the odds of a successful human genetic legacy.

Monogamy under the ancestral environment was also 'imposed' to a degree. The limited resources meant that any one man could not afford many wives, and in this sense monogamy was 'ecologically imposed' by existing the conditions and technological level. Many nomadic tribes are known to practice infanticide because the mother can only carry only one infant in her arms; again, were it otherwise, high fecundity would threaten the survival of the group.

That being said, human reproduction incurs particular costs. Only the female truly knows that the baby born is hers, as fertilization obviously occurs only within the female. Human reproduction might be contrasted to that of fish, where 'transparent reproduction' occurs externally: female lays eggs in the open, which are then fertilized externally by a male. Reproduction is 'transparent' in that the insemination process is visible to the two parties involved—in contrast to human

---

[112] As in birds, if every baby eaten was after reproduction, it would leave no genes for future and hence no genetic legacy.

reproduction. Inversely, as insemination occurs inside the female, it cannot be seen, and thus allowing for the possibility of cuckoldry with its consequent high cost to male. A male that takes care of progeny that is not his own incurs a tremendous lost investment, which accounts for the high rates of jealousy in human males.

Jealousy is a distinctly male trait, but obviously not exclusive to him. A study by David Buss had individuals imagine their partners sexually cheating on them. In males, the heart rate became extremely elevated, as if suddenly shot with 3 cups of coffee; the heart did not got back to normal softly after the event, as if a switch had been turned on. If we combine the ideas of Trivers with Hamilton, we are led to rather predictable results: the incidence of violence to women for cheating are extremely high. In the US, one third (33%) of all women are killed by their own partners, of whom 79.2% died at the hands of their current partner. In 2015 alone, 3,519 women perished as a result.

Crimes of passion have long history in human cultures, and have increased with the rise of 'easy kill' weapons as guns. Guns allow crimes of passion to easily occur, as the ease with which a man can pull a trigger does not impose a moment of reflection. These tragic assassinations are typically followed by the suicide of the killer, suggesting that the killer did not purposefully intend to kill their loved one, but rather had instinctively acted as if a switch had been suddenly turned on. Realizing the horror of their mistake, the killing of a loved one, they take their own lives as well. What is also particularly striking is that crimes of passion are broadly condoned. Until 1974 in Texas, for example, a man would not go to jail for killing a cheating wife. The lax legal stance is fairly common as all men can sympathize with the strong feelings of jealousy. Ninety-one percent of all men at some point have fantasized about the torture of a sexual rival.

Foolishly perhaps, females typically undermine the threat in which they place themselves. In this regard, marriage is an important institution for women as it provides the relationship with stability, which in turn tends to reduce the likelihood of violent events arising out of jealousy. Marriage does reveal particular and distinct emotional patterns, hinted at earlier. Males will tend to want to rush things and move faster in a relationship than woman, who in turn will want to go slower with regard to the maturation of the relationship. Prior to marriage male will be much more eager for a long lasting union, while females will show feelings of uncertainty and doubt. Could she do better? However, after marriage, pair bond formation occurs, and the couple will show strengthened bonds of affection immediately after the wedding. Pair bonding has an important genetic consequence. The bonds of mutual affection will be needed for effective child care and rearing, the birth of which will also serve to continue strengthening the pair bond—Darwin again being a good example. Darwin had a happy life-long marriage, due in part to his 10 children. By contrast, Charles Dickens divorced likely because he had had no kids. In contrast to marriage, divorce is particularly hard on women and their children.[113]

Note that often what spurs divorce is the clock of time. As males typically increase in status and income-earning potential, female lose their reproductive potential over the span of the years. Over the course of a marriage cycle, the interest between the two will begin to diverge, leading to the typical pattern of divorce. However, at the time of union between the two, it is in the strong genetic interest of the male to deceive not only his future partner but himself well.[114] Biology fools men into the sensation of eternal love so as to better fool women to believing the strength of their devotion, and thereby checking off a female's 'commitment criteria'. In the vast majority of cases, however, the claim is a lie—but an important one in that it allows for the continuation of the species. This is one way in which evolution encourages deception between individuals. As long as the union leads to a successful reproductive results, i.e. genetic progeny, then the lie achieved its ends. Evolution does not

---

[113] During the 1970s, the 'no-fault' divorce was established, allowing for a quick and clean break between the two parties. However, typically a male's net income increases. As a result of the divorce, he tends to experience an increase in his standard of living, and also to remarry. By contrast, female wealth sharply declines, and traditionally do not remarry.

[114] If one can fool oneself, and actually believe a lie (i.e. eternal love), one will much more easily be able to fool another.

necessarily care for human happiness at all, but rather it cares for is the passing of genes onto posterity. As Dawkins notes, one should not presume that one's own interest and those of one's genes are one and the same. While most of the time the two converge, they do not in every instance.

Tragically, there is an unforeseen cost to divorce. If a woman remarries, her children are much more likely to be abused by her future partner. Statistic after statistic reveals that children do not fare well at all under these circumstances. This is easy to account for from the standpoint of sociobiology, and one might equate it to a type of cuckoldry. Given that children are cost intensive, and the particular stepchildren are not the spouse's actual genetic children, the stepfather is losing a substantial investment.[115] It is here to be noted that after a rape, a husband's feeling for his wife and children also typically declines in what is a similar profile of cuckoldry—in spite of awareness of the victimization of the innocent party. This obviously constitutes a double blow for the woman.

Rape itself has been reinterpreted by sociobiology, and typically occurs under a particular set of circumstances. Feminist theory used to argue that rape was an act of misogyny—a hate crime against women. However, this interpretation does not accord with the actual data. The incidence of rape closely correlates with a woman's fertility, which peaks between the ages of 20 to 22, but then rapidly collapses after the age of 40. Rapists also have a typical profile. These are typically from a lower social strata, have low self esteem, and generally do not have many reproductive opportunities. Their reproductive capability is limited given their social circumstances, and they typically undertake a crude cost benefit analysis. Many rape victims do in fact become pregnant, suggesting that evolution could 'select' for such behavior as the rapists' genes are ultimately passed onto future generations. The actual social circumstances of rape are an important predictor of the activity. If there are few consequences, and the opportunity for the illicit act presents itself, there is a higher likelihood of its occurrence, as typically seen during times

---

[115] The tendency is diminished if the couple have children of their own.

of war. Rape is essentially an unforeseen mating opportunity for the rapist—suggesting a few principles young females should take to avoid becoming victimized.[116]

That being said, the key reason rape is so detrimental from an evolutionary perspective is that it eliminates female agency in the process. Because female victims of rape do not select for the rapist as a mating partner, it represents the counterpart to male jealousy. In both cases, there is a lack of voluntary and purposeful behavior, while at the same time constituting an imposition of the heavy costs of such activity. In other words, in the act of rape the genetic outcome is imposed by an external agent, not unlike the case of male cuckoldry.

Incidentally, polyandry or the feminine counterpart to polygamy, is extremely rare in human societies.[117] The practice has been observed in Tibetan villages, but only under particular conditions. Typically, a woman's multiple husbands tend to share very similar genes in that they are usually brothers, thus constitutes an example of inclusive fitness. Also, these tend to occur in rather isolated communities, where a younger brother has few, if any, opportunities for financial or sexual advancement. These liaisons are typically undertaken to preserve wealth within a family, as the abundance of children typically leads to an the splitting the family fortune. However, these arraignments tend to be highly problematic and very unstable. In essence, the younger brother acts as an eunuch, without any sexual privileges, and becomes a minor partner lacking any leadership or true decision-making in the family. It has been observed that if the younger brother's opportunities improve, the association usually dissolves. If the younger brother seeks greater independence or sexual rights, he will introduce tension into the tenuous arrangement. There can be no doubt that polyandry is

---

[116] Young females should not provide opportunity for the activity: never walk alone in dark and when possible always walk in public, for example. After the age of 40, the potential for becoming victimized will the sharply curtailed.

[117] While in *polygamy* a male has many wives, in *polyandry* one female has many husbands. The more abstract and general term is *polygyny* or one spouse to many partners.

basically a form of primogeniture; here primogeniture is masked as polyandry. [118]

What are some of the social implications of sociobiology? Republican conservatives typically pretend to defend 'family values' while also opposing any government intervention. Preferring to keep the inequality of the status quo, they also oppose all policies pertaining to the social redistribution of wealth—which from a strict genetic point of view is actually counterproductive to the creation of families and to the reduction of crime. In this, sociobiology reveals deep and unapparent internal contradictions of many ideologies.

When we turn to the richest nation in the world during the twentieth century, one of its most striking features is the pervasiveness of monogamy in the United States. Yet why would monogamy persist when is no longer ecologically required? The wealth of that nation would suggest a more common presence of polygamy than monogamy—which in fact exists, but in an altered form known as 'serial monogamy'. For example, Johnny Carson had a series of multiple young wives, which in turn were taken out of family market due to his monopolization of their peak reproductive years. [119]

It has been shown that polygamy as much as income inequality is the sure path to violence. Polygamous societies tend to be highly violent, the reason for which is fairly simple. Polygamy implies an uneven distribution of sexual resources; not all men can marry and procreate it. One might also suggest that there is no faster way to create a revolution, political instability, or a change of the political status quo, than to promote polygamy. It is in the interest of elite to have monogamy, whose 'pacifying influence of marriage' is well known. Polygamous men, or

---

[118] During the medieval period this social dilemma was resolved it through the right of primogeniture, where the eldest male inherited all and all other males had to make do, resulting in a very difficult life for the latter.

[119] For every young wife Carson took, another man is not able to reproduce and have a family. Donald Trump might be considered another example. Trump married young women during their peak reproductive years, successfully reproducing with each, before moving onto other women who were also in their peak reproductive years as his former wives aged.

leaders with many resources, by necessity also have to be violent. One Zulu warrior had 100 wives, and was so violent that sneezing at a dinner table could be cause of assassination. Polygamous men are under constant threat by rivals, revealing the costs to being a hawk than a dove.[120]

In human societies, rank typically correlates with reproductive success, due to the social nature in which resources are measured. This again helps to account for the relative high amount of violence in lower class ghettos. The relative scarcity of resources means that social status is a valued asset as it serves as an indirect determinant of actual and future resources (i.e. wealth). As in the case of the Zulu warrior, small slights as the manner of looking at a person might lead to a gun fight in a bar. These conflicts between young lower class males are often not over concrete discrepancies, but rather should be interpreted as fights over status, and in turn over potential material dividends and social resources accruing from this status.[121]

Low class status also severely impacts reproduction in another manner. Because there are few resources, males under these circumstances are likely to view crime as the only means to success. As such, crime might be said to be 'adaptive' with regard to reproductive success. From a fundamental 'point of view of the genes', the best chance to pass one's genes on is to enter into criminal activity—in spite of the social repercussions this might have over the long run. The large quantity of violence in '*puntos de drogas*' is basically a fight for reproductive success. In evolution, reproductive fitness rather than longevity is the main goal. As long as an individual reproduces prior to death, the individual has had genetic success, by definition. Thus, from a 'genetic point of view', the short and brutish span of a criminal

---

[120] Leaders have to expend a great deal of time warding off threats, which can be costly, resulting in a social state of one against all described by Hobbes. Doves, on the other hand, avoid entering battle, and hence can better use resources. They can, for example, can spend more time innovating and producing items which will contribute to the growth of the economy, instead of fighting one another over a fixed set of resources.

[121] Unmarried men tend to incur in risky behavior; for which, they are also more likely to land in jail.

life is irrelevant. Again, we cannot assume that genetic success is the same thing as happiness.

It could be argued that monogamy is an implicit contract of equality between males, rather than a pact of equality between females, which would certainly look very different.

Tit-for-tat and the origins of ethics

Knowing what we know of sociobiology, what principles and practices should we use to ground an ethics? What would a realistic ethics look like on this basis? Inversely, what is the ideal ethical state one can hope to expect; what is the tide towards which all should move? If men are not saints, can we only expect a reduction of extremes or is the entire project ultimately a futile one?

In order to appropriately answer these issues we must remember a key distinction. Humanity has evolved in two different contexts: that of the natural world.[122] and that of the social world[123]. As much as the natural context, the social context in which human evolution occurred molded its current form. The key to creating a realistic ethics arises from a consideration of both of these. The first points to the lower setting or the 'minima' (natural world) while the latter points to the upper setting or the 'maxima' of our expectations.

One should also never forget that all social relations are characterized by conflicts of interest. No two parties share exactly the same interests, and will diverge in some manner or other, to varying degrees. These conflicts may be minor or they may be very substantial. One might, for example, presume that a mother and her baby are as one given their usually tight bond, but there is a continual 'arms race' betweem the two, not unlike that which exists between the male father and the female mother. The baby is always trying to draw more resources from the mother than that she can provide, resulting in what is known as gestational diabetes.[124] It is interesting, however, to note the

---

[122] Flora, fauna, geology, geography
[123] Family, enemies and friends, community
[124] The fetus releases hormones which elevate sugar levels which is beneficial to it, followed by a consequent spike in insulin production.

benefits of the birth for mother. Her IQ typically increases by 23 percentage points as a result of her pregnancy. This makes sense, given that any changes which directly benefit her will contribute to her baby's survival, and here the genetic interests of the two coincide. The baby represents the genetic future of mother, and in its success, the 'fitness' test is passed. That being said, there is no denying that a conflict exists in spite of their common interest—and will persist for many years afterwards. One can also look at the case blackbirds.[125]

Conflicts of interest characterize all social drama, and stands as a core theme of all important tragedies. Similarly, conflicts of interest are a fundamental trait of all ethical dilemmas, be it a conflict between individual and society or one between selfishness and altruism. A researcher, for example, might be told to lie about the results of an experiment. Should they comply with this clear breach of professional code of conduct? In theory, one might presume that the researcher should not comply, as it is a violation of the implicit social trust of science broadly speaking—akin to a doctor that violates the Hippocratic oath of doing no harm. However, a number of issues might be at stake. The corporation providing funds for the researcher might need a positive outcome as it is about to go bankrupt. If it can hit a 'killer product' (drug), it will generate billions of dollars in revenue, and in turn help insure the employment of the scientist. The scientist himself might be threatened, and his own employment might depend on the outcome of the experiment. After all he also has bills to pay, a house mortgage he owes, and a family to take care of. Does he forsake these for abstract moral

---

[125] If one observes their interactions, one can often see a young blackbird squawking with its large beak open at another which is apparent its mother, asking for food, while at the same time insects surround it on the ground. In spite of its persistent requests, the young blackbird is physically as large as its mother, and hence does not lack the physical capacity to sustain itself independently. The mother might feed it now and then, which obviously runs against the mother's self interests as her own caloric consumption is reduced. At some point, the mother blackbird is eventually forced to stop, out of sheer self interest. Many human adolescents also reflect a pattern of immature behavior that will persist through early adulthood as well.

principles of justice or does he 'capitulate' to commercial interests?

What criteria should we use to decide? Do we try to make a rational calculation based on an inference of cost and benefit, trying to predict the damage incurred relative to the benefit gained? The poem by Robert Frost "Road not Taken" is poignant. ' I looked to end of road as far as I could see, paths diverged; which route should I take?' All key decisions in life share these features: a great deal of uncertainty, not enough information, and many significant implications. Momentous decisions can have a huge impact, and determine the course of one's own future life. Or do we try to establish decision-making on the basis of principle, abiding by certain values which are held so dearly that we are unwilling to compromise them.

Thirdly, do we try to make a decision on the basis of historical antecedent, using precedent to guide us in our moral choices? We might recollect the history of a particular scientist and remember the decisions he made under such conditions. The fame and renown of the scientist suggests that one should follow his example, and hence we make a decision based on his example, whatever that example might be. Do we try to assess the impact of our actions on our own reputations? It might be noted that corporate misdeeds are known to both the individual and to corporate representatives who promoted them in the first place, which in turn can always ultimately be used against the very scientist in an implicit form of blackmail. Under this scenario, wrongdoings can enter into a vicious cycle, spiraling downwards to atrocious ends.

Sociobiology will hopefully help answer these ethical dilemmas, by establishing firm criteria of analysis. It can help improve the ethical behavior of scientific practice, no matter what field that scientific practice may be: physics, chemistry, astronomy, and biology.

Let us first turn to the minima, the context of nature.

If George Williams and William Hamilton are correct, then the emergence of human societies is very odd and unusual. Nearly all societies today are composed of unrelated individuals. The degree of relatedness between any two random persons is trivial—and yet societies do exist. Society is the predominant unit in which humans live and interact with the world; so much so, that its scale increases year after year. The movement to urban areas has increased over the last century at a global level. The United States during the 1940s saw a rural to urban migration, particularly of African-Americans from the from rural south to the urban north and cities as Detroit and Chicago, resulting in the creation of jazz as a new cultural phenomena. Europe has also been characterized by a migration from former colonies to the metropolis, as from India to England. A good example of this is the case of the renown mathematician genius Srinivasa Ramanjuan, who studied under G. H. Hardy early in the twentieth century.

A question we must then ask is why are cosmopolitan urban areas so predominant throughout the globe? Is sociobiology wrong? Irrespective of Geoffrey West's studies, the amount of violence might suggest that sociobiology was right on the mark.

More than 80 million persons have died during the wars of the twentieth century. This figure is oddly small when compare to disease; the 1918 Spanish flu killed more than 50 million. But it is still a significant number. Note that 60% of these violent deaths have been those of noncombatant civilians. In other words, most who die are not directly involved in military conflicts, but are innocent 'casualties of war'. Steven Pinker has noted that the severity of war has declined; less individuals die in wars to day than they did in the past. However, this observation does not invalidate our prior claim.

What is particularly striking of human violence is that mammals of the same species do not generally kill each other. There is a built-in aversion to intraspecies predation. While there undoubtedly exists aggression during mating seasons, these are generally not life threatening. Typically a demonstration of submission is made when two dogs snarl at each other, and a dominance hierarchy is quickly established—which reduces the need for actual physical aggression.[126] This dynamic occurs in

---

[126] The lesser dog cowers by placing its tail between its legs. It shows its lower status and then recedes.

fights over food, reproduction and so forth, in what is often referred to as a 'pecking order'. In conflicts which follow therefrom, rank is quickly established by gesture.[127] In this dynamic, the need for dangerous violence with potential body harm and injury is drastically reduced, and is a trait found in all animals with some degree of sociality.

It is also striking that wartime atrocities committed by 'normal men', as noted by the social philosopher Hannah Ardent. She was struck that nonpathological family men made up majority of the Nazi army—the very same which committed the gross atrocities during WWII. It is curious to point out the particular psychological mechanisms of violent human predation. Humanity is basically not made to hunt itself, and hence during violent periods this sympathetic cognitive process is turned off. Mental traits show during these periods include the turning of potential targets and victims into symbolic 'animals', defined for example as a dangerous predators that needs to be killed for survival. The notion of 'turning bezerk' originated in Viking Norway, and implied of into an irrational 'bear' to fight. Another psychological dynamic, in the reverse, is that of turning the victim into a prey animal that needs to be hunted and killed for survival. Many examples have been drawn from the WWII literature, as the notion of shooting Japanese soldiers as if one were shooting country rabbits.

A third psychological technique is that of symbolically turning the victim into a disease—a move of which Hitler was particularly fond. He alluded to Jews as being akin to infectious disease that needed to be wiped out; as with personal hygiene, where invisible bacteria are warded off. The association to nonvisible bacteria or viruses implies that attacks on their persona and collective can never end are they are 'never enough' to defeat the plague. This cognitive framing pushes actors to extremes of genocide where the universal animal notion of intra-species submission is cognitively subverted.

---

[127] Instead of a 'fight to the death', animals 'peck'—a trait that was first observed in chickens (and hence the name 'pecking order').

It is striking how evolutionary past emerges in these wartime examples of human cognition: the projection of animal world onto the human world was imbued with traits which allowed for a shocking amount of violence against the same species, which is naturally curbed in other animals. Meat eating helps explain this behavior.

Meat occupies a universal role in nearly all human societies, as meat is shared in 'feasts'. The are a distinctly collective activity, of a public character which occur 'outside' and in plain view before all to see—as opposed to private activities obscured from public view. Typically all in members of the group are invited. Nonrelated biological members also participate in that meat is shared with those outside the group. Interestingly, chimpanzees also demonstrate the same behavior, which makes sense as chimps are our closest primate relative, sharing 99.9% of our genes—and one of the most violent as well. However, in contrast to the average human diet with 20% meat consumption, only 5% of the chimpanzee diet is so constituted.[128]

These meat rituals or 'feasts' have particular social benefits. In them, there is an abolition of all hierarchies. In chimpanzee communalities, it is to be noted that immediately prior to such feasts, there is a strong reinforcement of hierarchy.[129] There are subtle hierarchies associated with feasts.

Meat providers often have a much larger number of sexual partners, in both chimp and human communities. This phenomenon is known as *mate provisioning*. It would then follow that societies that are more promiscuous tend to have a higher incidence of hunting; a substantially larger amount of time is spent by males seeking meat. We may also observe modern

---

[128] The portion of meat is much higher in some particular cases, as with the Eskimo.

[129] Primitive human communities used to be idealized as egalitarian, and have also been characterized as glorifying nature—a claim which is somewhat exaggerated when one considers the a great amount of animal losses during indigenous buffalos hunt in the Americas. So many were killed when driven over a high cliff, that only those specimens on the outside of the enormous heap would be eaten; the vast majority would simply spoil.

dating rituals, key of which is a relatively expensive dinner; most of these are characterized by the eating meat.

One reason for the communal character of feasts is simply that hunting's success is usually the outcome of a great deal of luck. Typically, only 40% of hunters will be successful in their hunt, which consists of large prey and whose meat will quickly spoil if not eaten relatively quickly. Meat is highly nutritious, with a high caloric content that usually requires a relatively low amount of effort for humans to obtain. These facts help account for the predominance of the behavior. A hunter at any point might need to become a beneficiary, and thus its makes sense for him to share his spoils during a time of plenty following a successful hunt; there is a low cost but a substantial long-term benefit to the practice of reciprocity. These dynamics were captured by George Catlin, a painter who traveled during the nineteenth century throughout the wild west in the United States to capture indigenous life prior to the arrival of Western Civilization. Aside from noting that native Americans believed paintings literally 'captured their souls', he also observed that food sharing was fairly common—particularly in the case of an Indian traveling alone and starving.

What implications does this have for the ethics of biology and animal experimentation? Can we save Beatrix Potter's *Peter the Rabbit* and all of the fictional animal characters from beloved children's stories?

Sociobiological theory suggests that animal experimentation is not likely to end any time soon, as it is too similar with hunting. Humans have no direct genetic link to animals, broadly speaking and hence there is no direct genetic benefit to the absence of animal experimentation; quite the contrary. It also suggests that meat eating is also not going to end soon, as it is an evolutionary adaptive phenomenon closely tied to the total population. If humans exist, they will always tend to eat meat of some sort or another; the greater the population, the greater the meat consumption. In fact, its limiting capacity is actually defined by national wealth. China's meat consumption during the twentieth century clearly reveals that the percentage of meat consumed increases proportionally to the increase in national wealth. We should not have too high of an expectation with regard to animal experimentation and consumption, as these trends will likely continue long into the future.

Let us now turn to the maxima of social life.

After WWI, Albert Einstein asked Sigmund Freud to provide a psychological explanation for the war. Freud by then was an old man, but took up challenge nonetheless. Einstein personally believed war to have been irrational, and hence believed that a psychological study of the unconscious might be able to elucidate it. Freud provided his answer in the book *Civilization and Its Discontents* (1930). In it, he argued that sexual urges were not the only human motivator, but that there was an internal self destructive one as well; the counterpart to outward aggression was a self-directed inward aggression. This notion we know to be incorrect.

Behavior does not exist in a vacuum. Freud ironically saw himself as a Darwinist and evolutionist but in fact was not, as he believed in Lamarckism, or the inheritance of acquired traits. There are some general similarities between Freudian psychiatry and sociobiology. The principal concept of the former are characterized the notion of unconscious drives which conflict with the conscious mind; id, ego, and superego matches to a degree the reptilian, pre-mammalian, and the 'neomammalian' brain. Freud criticized that too much importance and power was given to the ego. Sociobiology (evolutionary psychology) makes similar distinctions in that the mind is not always conscious of its actual motives and is rather very good at rationalizations.[130]

There are, however, many problems with the Freudian interpretations of the psyche, the most obvious of which is perhaps the Oedipus complex. In it, the son is in love with his mother, which creates an eternal tension with his father. However, from an evolutionary standpoint, the claim makes no sense, and in fact the cases of mother-son incest are extremely

---

[130] This feature as been well demonstrated in mirror experiments, where each eye sees a different image. Individuals who for some reason have had their corpus callosum severed, will provide rationalizations based on a word shown only in one eye; they did not recognize that they ever saw the word in the first place—in spite of its enormous influence.

rare. The explanation for this is relatively simple: the genetic costs of incest are onerously high for mother, who will bear the direct burden of all physical deformities which will inevitably rise. However, the son-father tension can be better explained on another sociobiological basis: both compete for the mother's attention. That is to say, the father wants to procreate, while it is in the son's genetic interest to prevent the arrival of a sibling, and also seeks to monopolize parental resources for himself. In polygamous society, as son passes pubescence and turns into man, he actually competes with the father over other women in the group, but never for the mother properly speaking.

While Freudian psychiatry can provide an interpretation for some behaviors, its explanatory schema is rather circumscribed and problematic.[131] In this, sociobiology provides much better explanations for human behavior than its counterpart. Because humans have lived in social groups, the notion of reciprocity is critical to understanding all human interactions—in particular the notion of reciprocal altruism 'tit-for-tat' developed by Trivers. Reciprocity has been universally encoded and can be found in a number of different cultures across history, amounting to the notion of an eye for an eye, tooth for tooth. If someone helps you, you help them back; inversely, if someone cheats you, you do not help them. The dynamic is key in the establishment of alliances or 'friends', which strengthen the status of the individual and provide assistive support in times of need.

Humans, in other words, have a unique sensitivity to cheaters.

The brain is a differential engine which can quickly and easily detect inequalities of exchange, so much so that it has become one of the brain's underlying cognitive biases. As noted by Matt Ridley, such propensities aided human survival, and is well illustrated by various examples where the ease or difficulty of detection is used as an evolutionary cognitive marker. Closely associated to it is the notion of reputation and public information generally speaking. To be recognized by one's community as 'kind' or 'good' has many social benefits, both direct and

---

[131] Curiously, Freud tended to have a big ego himself, which can be contrasted to Darwin who was excessively humble, to a fault.

indirect. Individuals of high status or resources are much more amenable to sharing those resources, as such aid comes at very little cost to themselves. A poor reputation for generosity in tribal context can be deadly if driven to the margins of the community. Historically, being shunned from the group placed one's life at risk by drastically increasing the likelihood of being killed in precivilzied world. Contrary to common opinion, the premodern world was anything but pastoral.

More importantly, notions of human justice emerge directly out of evolutionary reciprocal altruism. The notion of fairness is innate in primates, perhaps best illustrated in Franz van der Vaal and Sara Brosnan's capuchin experiment of 2003.[132] The human sense of justice and fairness has been evolutionary 'built in' to our psyche millions of year ago, as it is shared with a common ancestor. This in turn implies that our human sense of justice will not disappear any time soon, and is universal across all human societies, regardless of how it is actually put into practice. It is from this evolutionary ancestry that our sense of ethics and morality emerges.

A computer study by Robert Axelrod showed how it could easily emerge. Axelrod requested that programmers build different models of behavior, and did not place limitations with regard to length or complexity. Each computer model was then made to interact with each other in a series of exchanges totaling 200 encounters. Surprisingly to all involved, the program that end up winning was a simple 5 line code 'tit-for-tat' program written by Anatol Rapoport. The initial encounter was positive, in that Rapoport programmed his model to offer something to its counterpart in its first exchange. If the other did not return the favor, then it would reject a future exchange with that particular program. If, on the other hand, the other reciprocated, Rapoport's

---

[132] In the experiment, quid-pro-quo exchanges were established between the investigators and the capuchin monkeys. For a given task, each monkey was given a particular 'payment', either a cucumber or a grape. Capuchin monkeys love grapes more than cucumber. When these were placed next to each other using glass cages. When one observed that for the same task, they were only given a cucumber while another monkey obtained a grape, they would immediately complain and 'go on strike' by refusing to perform the task.

program would then return the favor. By the end of nearly all the runs, Rapoport's program ended up winning.

However, there were various conditions to its success. If the number of visitors increased drastically, quickly entering and leaving the exchange, the tit-for-tat model would not win. The reason for the outcome is that no history was generated for the model to learn from; as every visitor entered in and out of the game, no history or long term experience was ever gained by it, and (more importantly), a trading partner's reputation could not be established. This experiment showed the role anonymity plays in the formation of crime in modern urban areas. Similarly, if there was only one tit-for-tat program who existed in a sea of aggressive programs, it would also be swamped out. As all exchanges were negative, there was no net beneficial result from being good at the beginning of the encounter. However, if a group of tit for tat entered into the scene, the small group would ultimately be able to modify situation to a more successful outcome. It was clear that successful behavior is conditionally dependant, rather than being 'absolute' and 'a-historical' or 'a-contextual'.

Axelrod and William Hamilton ended up collaborating in a groundbreaking book, *The Evolution of Cooperation* in game theory.[133] Prior to their collaboration, it was typically assumed that the strategy of defection always won in a prisoner's dilemma scenario.[134]

---

[133] Rather ironically, it came out in 1984, the same year of George Orwell's magnum opus of an totalitarian dictatorship model.

[134] The prisoner's dilemma consists of two prisoners, held in separate rooms and whom cannot talk to the other or have any information of the other. Each prisoner has the same dilemma: do you betray a partner and get a reduced sentence by sending the other to jail or do both partners say nothing, and in the process mutually win. Prior theory suggested that in the majority of cases the 'defection strategy' was the optimal for each prisoner. It was generally best to defect first, before other one did, and is a strategy typically seen in criminal prosecution cases—as that of Michael Cohen and Donald Trump. The minor player with enough information is offered a deal in exchange for his testimony against his major partner; defection wins, as there is little to gain by not stating anything given the great deal of uncertainty of the scenario. Players do not know if the other players will defect as well; criminals are not known for their

There was an important implication of Axelrod's study. Reputation is key in all social behavior, and the reason why so much emphasis is placed on it. It was also clear that the role of information is equally important, particularly the role of gossip in human communities.[135] A large number of studies from various fields support this conclusion. Sociological studies of crime have identified its causes, anonymity being one of the principal causes of crime.[136] The notion of moral hazard is also relevant; upon the absence of consequences and impact on public reputation, individuals have a much higher probability of committing a crime; criminals are prone to hiding their faces for similar reasons.

Historical studies also illustrate the veracity of the observation. Nineteenth century England was characterized by a Victorian mentality, whereby a great emphasis was placed on decent behavior. Such high standards of conduct were established, that they now seem a bit extreme from our contemporary point of view. However, cities at the time were really relatively large small towns, where everyone knew each other and thus a good or bad reputation could spread quickly. Hence public behavior was severely restrained, in light of the opinion of others—a good example of which is Darwin, once again.

Darwin lived and worked in Shrewsbury, where everyone knew each other. He became known for his ability in making friends, and even started a social security association in the town—a form of assistance in case of emergency. Darwin even built a garden for the Patagonian Fuegians in England, who would actually never be able to reciprocate the favor. However, Darwin's case clearly reveals the long 'reach' of reputation, with a solid reputation as being a 'good guy'.

---

high moral standards—for which they would otherwise not be criminals in the first place.

[135] Prior theory already pointed that a player who knew his partner had not defected, would not himself defect.

[136] It is psychologically much easier to commit a crime if you do not know the person.

Psychological studies of peer pressure also support sociobiology's contentions of the role reciprocal altruism in human behavior. In a famous experiment, ten individuals were placed in a room; of the ten, only one was the actual subject of the experiment.[137] The principal subject was asked to select the smallest stick out of a stack of six. While it was obvious that 'stick B' was the shortest, all in the room except the test subject picked 'C' as the smallest—a claim which was patently false. Being aware of the consensus opinion in the room strongly influenced the test subject's assessment. In a majority of cases, the test subject selected the wrong stick when asked to provide a final answer; these sought to go along with the group rather than follow the overwhelming evidence before their eyes.

Human brains are differential engines that are very finely tuned to detect cheats, so much so that even schizophrenic patients with damaged brain that can see or hear nonexistent entities, will be able to easily detect cheats in spite of their damaged brains. In spite of their power, however, human cognition has a tragically difficult time identifying altruists, so well described by Ridley.

Reciprocity and gift-giving are thus a universal features that cut across all human societies. We have previously alluded to this phenomenon with regard to meat. All societies have feasts where meat is shared; upon the receiving of a gift, there is the expectation of reciprocity at a future date. Failure to do so only demonstrates that the individuals will not abide by the rules, but also show oneself to be untrustworthy. This social dynamic was taken to absurd heights by indigenous tribes of the Pacific northwest. Great celebrations would be held where gave costly gifts were strangely given to moral enemies. While it might be noted that it was better to give gift to enemy that to kill them, the gift is actually being used as a form of manipulation, so as to better control other individuals. A sense of debt was established between the two parties by the ritual. Such tribes are not the only ones who have taken advantage of this human trait.

---

[137] All others were assistants to the experimenter.

Similarly, the concern with reputation is another universal human constant. We all deeply care about how others think about us at some deep psychological level. Again, the case of Darwin is a good example. He was very conscientious and purposefully sought to continually build alliances. Note how cognition is influenced by these dynamics. We tend to esteem and value individuals which will be of clear benefit to us, be this benefit direct or indirect. The role of status thus represents a positive bias towards a favorable treatment of the high status individual, or another cognitive bias influenced by evolution. Social status or the esteem of others is not something you can buy or barter, but something that must be earned. Again, Darwin is a good example.

Darwin was first introduced to the naturalist John Henslow by his cousin William Fox. In his letters, Darwin is initially very apologetic to Fox, and deferentially views Henslow as the best scientist of England. However, upon Darwin's return from South America and the consequent fame he acquired, Darwin's relationship with Fox was reversed. Fox was now very deferential to Darwin. Similarly, as he matured as a scientist, Darwin became good friends with Charles Lyell and Joseph Hooker, the latter of which became integral to the defense of Darwinian evolution at the Linnean Society debate. Darwin often used Hooker as a sounding board, somewhat akin to Einstein and his friend Besso. Lyell also became a very good friend of Darwin's, in part for their mutual assistance; Darwin provided a lot of evidence for Lyell's scientific ideas. Initially, Darwin idealized Lyell. However, as Darwin became the famous "Darwin", he became much more critical of Lyell. He began to note, for example, that Lyell tended to seek the approval of the powerful and the wealthy; Darwin's social ascent meant that he was now much more willing to see flaws which he previously ignored.

There can be no doubt, however, that both of Darwin's friends were ultimately critical in overcoming his first major crises: Alfred Russell Wallace's paper on natural selection. Today natural selection is known as Darwin's theory rather than Wallace's precisely due to the assistance of Darwin's friends. If Wallace had sent his paper directly to a scientific journal rather than to Darwin for verification, the theory of evolution would have acquired a very different history. Darwin pretense of magnanimity by forwarding Wallace's paper to a neutral British scientific journal was a false one, fooling Wallace all along. Today Wallace, unfortunately, has become but an appendix in the theory of evolution.[138]

What then, is the most effective 'ethical policy' that should be adopted based on the findings of sociobiology? Given that the social context is the predominant one in human affairs it suggests that a core component of any ethical policy will be an emphasis on reputation. Reputation is the principal coin of value of the scientist. A good reputation amounts to high social status, and all the consequent goods that are associated with it. The implied higher level of trust means greater access to funding, a greater number of colleagues, recognition of scientific merit, of a much higher likelihood of being read in the first place, and so forth.

Again, reputation is a universal feature of all communities, be they human or primate. Again, we may look at the case of Darwin. People were often amazed to see how zealously Darwin guarded his reputation. He was always very careful in his expressions, and taught his son to write deferential letters. As we have seen, this reputation building served Darwin well during his career, specifically during a potential crises upon the arrival of Alfred Russell Wallace's manuscript. Hooker and Lyell did Darwin a huge favor; had Darwin's enemies been in their official positions, the outcome would have obviously been very different from what it was. As Wallace today, Darwin might have ended a footnote in history, tragic when on considers the amount of time he had spent working on the theory.

From a personal point of view, it thus behooves the scientist to not only be ethical, but to also appear to be so. As Jose Arsenio Torres once noted, the queen must not only be modest but must

---

[138] Darwin had spent ten years working on barnacles, hopefully obtaining solidifying evidence of this theory. Darwin made the wise move of pretending to pass Wallace's paper onto more important fellows, and Wallace incorrectly believed was being helped, not realizing the close relationship of those involved. As is now well known, it was jointly decided to publish Wallace and Darwin's work on the topic, without informing Wallace.

appear to be virtuous as well.[139] Consider a counterexample. If reputation is the coin of the scientist, and perhaps of all academics, then violations can be easily be used as blackmail. In other words, a violation or breach of ethical conduct can result in a vicious spiral used to manipulate the individual to ends that may not be theirs. Again, remember that the problem with both jealousy and rape is that the outcome is not determined by the individual but rather by an external agent. The blackmail of a scientist might consist of following: if you do not do 'X', we will make your personal ethical breach public. The severity of that personal ethical breach will greatly impact the effectiveness of the blackmail; the greater the ethical breach, the more effective the threat due to its potential personal repercussion.[140] In serious violations, the scientist thus easily becomes a puppet of an external agent to science, leading not only to the personal loss of his scientific integrity, but more importantly to the degradation of the scientific enterprise as a whole.

Generally speaking, any reduction of dependency and vulnerability by an individual is a good policy. Do not fall into unmanageable debts. Try to establish alliances with leaders in field; small gestures of kindness can have a big impact over the long run. The greater independence a scientist has, the greater the autonomy of speech he will carry. The independent scientist can act according to their conscience, and, more importantly, act against its violations. Benjamin Franklin once noted that liars are always easy to spot for their convoluted claims; the mind is not made to handle multiple truths to multiple individuals, and eventually will not faithfully recall his own lie—thus making it easy to identify. Ultimately, the loss of credibility is the greatest injury a scientist can suffer.[141]

---

[139] Arsenio Torres is a public scholar who once regularly appeared on the radio, now retired.

[140] The higher the likelihood of folding before the threat of blackmail.

[141] Example of this dynamic in Italy have actually been studied. Diego Gambetta's *Codes of the Underworld* (2009) is a fascinating analysis of the degradation of academic values, specifically the practice of meritocracy. "[I]n a corrupt academic market, being good at and interested in one's own research, by contrast, signals potential for developing one's career independent of

Sociobiology thus teaches us that an important principle required to establish ethical behavior is transparency. The tit-for-tat model specifically showed that the lack of information promotes injustices and the abuse of power. There are many example of this dynamic, as the Nazi concentration camps which were hidden from pubic view in Auschwitz, Treblinka, and so forth. A particularly noxious case of a father – daughter incest in Austria tragically illustrates the point. The case of 'Josef F' came to light in 2008. The father hid his own daughter in a basement for 24 years, from the age of 18 to 42; having three children with her. When one of the children escaped, the authorities were alerted of the crime. Although the daughter was finally freed, there can be no doubt regarding the psychological impact on her, whose reproductive years were confiscated by her own father in what is perhaps one of the most grotesque incidents of incest ever recorded.

On the other hand, positive examples also provide ample evidence. Again, given that small towns predominated in Victorian England, gossip was the principal currency of friendship. With abundant information, improper behavior would be ruinous to an individual, thus leading to the particularly rigid Victorian mentality with its own pathological traumas Freud enjoyed discovering. It is to be noted, however, that the rates of crime were very low in this setting; and consequently the investment required by the community for police and surveillance was also consequently very low. This might be contrasted to modern anonymous and urbanized America where high rates of criminality occurs.

As a general rule, the greater the transparency in an institution, the greater the possibility that bad deeds will become known and hence severely undermine the probability of unethical or criminal behavior. In this, transparency can take many forms, be it physical transparency using glass walls, organizational transparency using random unannounced inspections or Jeremy Bentham's *panopticon*, or informational transparency where the

---

corrupt reciprocity. This makes one feared." p. 44. The use of reciprocity as a social tool in Italy is very similar to that of the Pacific Northwest Indians.

press serves as a fourth power, letting the public know of internal misdeeds. There are specific tradeoffs, however, that should always be considered when implementing such a policy.

Privacy and secrecy are an integral part of innovation after all.

Aristotle and Religion

Any discussion of ethics inevitably has to include religion given that religious codes typically contextualize ethical decisions in a society. Four billion persons live under some variant of an 'Abrhamistic' religion (monotheism)—Judaism, Christianity, or Islam—or more than half the world's population. All religions have claims pertaining to bioethics, including the origins of life, the dignity of man, and the nature of evil.

There can be no doubt that the status of all religions have been affected by rise of science. Claims which at one point in world history could have unquestionably been made are today no longer accepted. Al Ghazzali's Islamic theology implied no rational coherent order to nature, and hence the absence of any scientific knowledge. If an arrow was shot, by the time it landed God could remake the world in an instant, he claimed. Prior to the rise of modern science, there was little rationality to moral-religious claims. Today, such a claim would fall on deaf ears.[142]

We might then ask what the interaction between science and religion has been, or more specifically inquire into the impact of evolutionary Darwinism on religious thought and the framing of ethical decisions. How have religions reacted to the challenge of science, and more specifically to the ethical implications of Darwinian biology?

We will look at the Catholic tradition so common in Puerto Rico, and note that such reactions have historically not been positive or productive as a rule. There have been improvements in the Church's stance during the twentieth century, if not out of

---

[142] Such irrationalism was unfortunately widely used to repress early Arabic protoscience during the Medieval period. All of a scholar's work might be burned or, worse, he would have to flee for life. Political changes in this context could have huge implications for a scholar and his life's intellectual efforts.

the gross contradictions between many religious claims and established scientific facts. Pope Francis recently claimed that 'hell does not exist' in an open recognition of absurdity of some traditional Catholic positions based on a Medieval worldview.

But it is as of yet to be determined whether such profound theological changes will be accepted. There exists a strong conservative element within the Catholic Church, which will inevitably oppose any and all efforts at religious renovation—such as the use of the Bible as a litmus test of belief and truth. Pope Francis might en up ultimately becoming expelled as Pope for his 'extreme' liberal views. Institutions as a rule confer a great deal of tantalizing power that is very hard to discard, and which obtain a life of their own—even when the original causes for their emergence disappeared long ago. Becoming a political force of their own, institutions continually seek to keep themselves relevant, even while the intellectual basis on which they stand have crumbled long ago.

To understand this dynamic, we have to first turn to Aristotle, who was used to establish the foundations of medieval Catholic scholasticism.

Although he is more well known for his contributions to logic and cosmology as this is the work that was most influential in Western history, Aristotle was primarily a 'biologist'. That being said, these should be regarded as the outcome of his biological world view, and it is where most of efforts placed. One fifth of his entire corpus is made up of biological works, and it is biology which establishes principal paradigm for his cosmology. The emphasis on the study of final causes, for example, is pervasive in his natural philosophy. For Aristotle, everything had a purpose and a finality, and the notion deeply influenced both his physics and astronomy. We might actually have to redefine the Scientific Revolution as ultimately being the establishment of a break from Aristotelian biology, in particular a break from the biological paradigm to create physic's own internally consistent dynamics.

Three of his volumes in biology were published, but are incomplete, and read more like a compilation of notes lacking internal coherence and consistency as pointed out by D'Arcy Thompson—an anomaly when considered in light of Aristotle's final treatises in other areas. The volumes include *Historia Animalum* (or Natural History), *De Partibus Animalum* (On the Parts of Animals), and *Generationae Animalum* (On the Generation of Animals). In the second volume, Aristotle makes a poetic appeal for the study of biology to the young Athenian elite.

Aristotle's work is principally a rejection of Presocratic thought. The Presocratics pretended that all explanation had to be based on a single cause, and in their debates postulated various universal causes: air, water, energy, and so forth. Aristotle's main criticism of the period's thinkers is that the identification of a substance's material did not constitute a complete explanation. On the contrary, one needed to understand the growth and development of an entity to truly understand it; to what ends and to what purposes did a biological structure serve? Ironically a similar error to that of the Presocratics still occurs today in biology as a form of modern reductionism present in the work of T. H. Morgan.[143] Aristotle thus made distinctions in the different types of causes so as to provide a wider range of explanations that were fairer to the objects under study. This helps account for his particular creation of four causes: material, formal, efficient, and final.

However, not all causes had the same weight, and for Aristotle, the ultimate cause or finality was the principal key to understanding the universe. We can see how deeply influenced he had been by his own biological studies, as it was particularly in biology where the particular explanandum was most appropriate. Living entities are complex arrangements of complex structures, or 'infinitudes within infinitudes' as Leibniz noted in another historical context (*monads*). To understand a muscle, one cannot just look at the flesh, but had to see it as part of larger component. The heart muscle played a particular role, for example; to understand it, one needed to place it in context, as an

---

[143] In the nineteenth century, two general camps in modern biology emerged. The first is a reducitionist camp, whereby all biology was reducible to chemistry and physics. It was predominated by experimentalism, laboratory biology. The second camp is the evolutionist, who in contrast study macro dynamics, focusing on functional aspects. TH Morgan, had a very despective attitude of the latter.

integral part of a broader system—an observation which is true of all biology. One needed to identify how each part played a particular role in the totality of the functional body. Knowledge only of its component materials—blood and bone, for example— was simply an incomplete biological explanation.

Incidentally, embryology was seen by Aristotle as a development to final form, in the adult form which would have specific features. It was to this final structure which the young grew. As a consequence, monsters were inversely defined by Aristotle as lacking the ability to achieve their final form or purpose. Since males gave shape to things and females provided the material base, the latter were accused preventing entities from establishing their final forms, and hence the cause of all monstrosities. This helps account for the low place to which Aristotle relegated females. Given that Aristotle defined heat as a key determinant of change in universe, one can understand the emerging value structure of his universal schema. Females tended to have colder body temperatures than men.[144]

Finality in biology is thus turned by Aristotle into the principal scientific paradigm of the universe, and applied it not just to biology but to cosmology and physics as well—a notion which is today referred to as 'teleological' or the identification of the final ends or purposes for which something exists. Motion was surprisingly couched in this philosophical framework, and served as an implicit anthropomorphization of the universe. "Natural motion" was that which fulfilled an entity's mix of elements, while "violent motion", as in the case of biological monstrosities, was an external force which prevented its fulfillment. As an object moved closer to its destined place, it accelerated in speed. As objects in a void would travel at infinite speeds, which for Aristotle was an impossibility, ergo there could not exist a void. Because the supralunar realm was composed of aether, whose natural circular motion was eternal and perfect, it could never be influenced by unnatural violent motion.

With its incorporation of Aristotelism into Catholic theology, the notion of final cause became key principle during the

---

[144] On average, males tend to have higher body temperatures.

Medieval period, profoundly shaping its cosmology and world view. Hell is at the center of the Earth because that was the most imperfect point in the universe; the Christian worldview was not just geocentric but diablocentric as well. The universe was a static universe in that there was no fundamental change in the world. There might be historical cycles from one to the other but broadly there was no long term change. The world was set at beginning, and all creatures of world composed a 'great chain of being', each holding its particular role in the universe. In this biological 'plenum', all animal structures were set, all intermediate shapes existed in a hierarchical suprastructure. The topmost form was man, the apex of God's creation in that he was created in the image of God himself and reflected His goodness. Man was uppermost in conscience, while the lesser creatures of world lacked reason.

Generally speaking, the Church did not like change, historical or otherwise, and was very hostile to notion of a dynamic universe. Theologians who argued for this position were quickly imprisoned for life and tortured, as the notion of change implied the imperfection of God. Regularly perfection of the world was used as rhetorical proof for the existence of God, or the idea of a divine maker. This was strongly postulated by John Ray and William Paley.[145] A typical example was that of the human eye, which was very sophisticated. Eyes are typically made of materials with diverse indexes of refraction, thus all wavelengths converge on a single point; God's design was perfect in that the blurriness of achromatic aberration was accounted for in His design.

The Christian cosmology, as the Aristotelism on which it was based, was extremely coherent. It was sophisticate mixture of parts meant that an attack on any single component did not necessarily undermine its entirety. One would need to replace the entire structure, and alternate physics, which is what ended up occurring during the now much debated Scientific Revolution. By providing an alternative view of cosmology and its

---

[145] Paley's work was very sophisticated, using centuries worth of biological data from the New World which had accumulated through the Colonial period.

consequent mechanism (Newtonian physics), the Scientific Revolution upended the coherence of the Christian world view. The notion of 'final purpose' would ultimately have to be abandoned.

That being said, the notion of finality within the field of biology did not become obsolete, given its usefulness to that science. This position was retained until the nineteenth century when the Darwinian revolution also upended this notion. It goes without saying that there were consequential implications for Catholic theology and morality as a result.

To understand the implications of the Darwinian revolution on ethics, one thus has to step back and look at prior arguments. Catholic morality was based principally on the notion of the dignity of man. To repeat, man was unique and special given that he was made in the image of God. Man had conscience and a mind, and a soul—a notion that emerges from Aristotelian accounts of human rationality. As a result of man's conscience and mind, moral codes applied to him, which in turn were embedded into social laws.

These views were particularly poignant with regard to the innocent. In the Catholic moral worldview, those who had committed no moral sin held special place, and were to be contrasted with sinful men, whom could not only be punished but killed as well for their evil deeds. Intellectuals who challenged traditional Catholic dogma view were included in this definition, as the infamous case of Giordano Bruno.[146]

These theological positions had ethical implication for particular acts, as those of euthanasia, infanticide, suicide. As a condition of their innocence, these acts were prohibited in Catholic ethics, as in the case of the moral elderly or of children who knew no sin. For St. Augustine of Hippo, man was special and unique, and argued that the sixth commandment "Though shall not kill" also prohibited the act of suicide. The generality of

---

[146] Giordano Bruno postulated infinite universes, which in turn raised key theological issues with regard to the trinity. He ended up being burned at the stake.

the phrase implied its extensibility to the self as well as to other human beings.

Catholic theology also treated the problem of evil in a similar fashion. If God was almighty and powerful, why did he allow presence of evil in the world? There could be no doubt that the suffering of humanity is immense: the death of loved ones, those incapacitated, maimed, or injured, or even the inability to live a fulfilling life suggested inherent limitations to Catholic dogma. God was neither all powerful nor all good.

However, there were various counter arguments refuting these position this within the Catholic Church.

It was argued, for example, that men were given free will, and as a condition they were allowed to do evil, which is what de facto happened. In this interpretation, evil was a way to understand and appreciate goodness, which one could not appreciate until it had been lost. Evil was also designed as a way of improving human character. The difficulties in our lives make us strong, and help us develop courage and compassion. Finally, evil was also defined as punishment for man's sins, specifically the problem of original sin of Adam and Eve's betrayal and expulsion from the garden. The development of conscience was ultimately a byproduct of evil, hence the necessary existence of evil for the emergence of humility in mankind.

The publication of *The Origin of Species* in 1859 had enormous religious implications, and one cannot help but observe how drastically different its worldview was when contrasted to Catholic dogma. Darwinism postulated a non static universe; the notion of fixity of species was incorrect; and change in the biological world was non directed and nonteleological. There was no finality and purpose to the universe. Worst of all, the traditional definition of evil could be legitimized within the basis of Darwinian theory.

When seen from a Darwinian viewpoint, the problem of evil emerged as an extremely limited and anthropocentric issue. The sheer number of animals that have suffered far exceeded the suffering of men. Ninety percent of all animals that have ever lived are now dead. Most of these died in tortuous ways, either as victims of other predators or noxious parasites as *Cymothoa exigua*, a parasitic louse which substitutes itself for the tongues of fish. The wasp paralyzes the caterpillar, to deposit its eggs in that living body. It was hard to believe this evil had any edifying role for man. Not just that, man was but a recent actor in the biological landscape.

Darwinism completely overturned the context of evil and its religious interpretation. The edification of evil was senseless and nonsensical when we consider that, in the vast majority of its instances, man was not there to receive its lessons. In this new context, the traditional argumentation made little sense, as if God had created an enormous theater for a nonexistent audience to receive its lessons. We may also notice that the traditional definition of evil is very limited in scope, as its focus had been on man. Again, the animal kingdom reveals an enormous scale of evil, orders of magnitude above those of human communities.

Worse still, the problem of evil made sense within Darwinian paradigm. 'Evil' was simply one of the mechanisms by which evolution occurred. Varieties are gradually weeded out simply because the vast majority are composed of errors and poor design. Most lines of biological descent are now non-existent. Not all morphological arrangements are equally valid, the criteria of which depend on their environmental context and influenced by the creature's prior biological form.

Darwinian theory also had profound implications for the treatment of animals. If one took Biblical principles literally, one would have to conclude that the traditional treatment of animals was utterly criminal. As 'innocents', animals were the most innocent of all, and yet were regularly used for food, work, energy, and materials as leather (cows), glue (horses), or musical strings (intestines). There could be no moral justification, formally speaking, for these atrocious acts. Yet justifications had been created by Descartes, based on the nature of consciousness, which in turn justified the brutal treatment of such innocents, as illustrated in the work of Nicholas Fontaine in 1738. Fontaine's *Memoire* describes his mechanical experimentation on the structure of dogs. Their paws were nailed onto a board, in spite of the shrieks of pain and suffering, and their tremendous efforts to be let loose. Fontaine even made fun of those who sought to defend the animals. In his mechanical world view, the body was simply a series of levers, pulleys, pipes, pistons, which he sought to account for in his research. It is perhaps no wonder that Michel de Montaigne became one of Descartes' strongest critics.

Darwin was well aware of these trends, taking a position akin to the Lockean stance previously described. If man shared a common heritage with animals, then the difference between the two was one of degree and not of kind. At one point, Darwin decided to forego language in his study of animals, which had formed a basis of Christian and Cartesian argumentation, and decided to focus only on their behaviors. In doing so, it became clear that animal behaviors were not that different from men's. Under torture men will scream and will also seek to remove themselves from source of pain, as the many animals Fontaine experimented on. The faces of both showed the suffering of pain.

His own study of earthworms, *The Formation of Vegetable Mould through the Action of Worms* (1881), was particularly surprising to Darwin. It had tremendous implications for both agriculture and ethics. He noted, for example, that the 23 inches of topsoil was the direct byproduct of earthworms, which produced nutrient rich soil by using it for their own nutrition. This observation is a key trope of Darwinism: momentous changes were result of the accumulation of the small actions of countless individuals over a long time period. The most apparently trivial of creatures were the most important for human survival.

A second feature of the study was perhaps the most surprising to him: worms had intelligence. He had noticed that worms gathered leaves into their burrows by the stems, and wondered whether this behavior was instinctual or learned. He looked a various alternate hypothesis, as instinct and trial-and-error. In his experiment he removed all local leaves, and instead substituted these for exotic leaves of different and irregular forms the worms had never experienced before. To his complete surprise, the worms modified their behavior, and again took the differently shaped leaves by the stems, drawing these into their burrows. Not all animals consciously modified their behaviors, however.

From these experiments, Darwin drew the notion of an emergent conscience via evolution: morality emerged directly from the sociality of creatures, which was influenced by an animal's level of intelligence. Intelligence led to different levels of understanding of cause-effect relations, and hence to a varied moral sense. Because primitive tribesmen could not understand the long term impact of their actions, these persisted in them, as the drinking of alcohol which was not associated with the bodily harm this activity produced. As a result, different moralities emerged over time relative to the respective scientific maturation of diverse societies.

There was no doubt in Darwin's mind that a moral conscience emerged within a social context. Individual behavior is molded by the conflicts between immediate drives and long term social pressures. Reflection of an immediate selfish action would lead to regret, and in turn to a modification of an individual's behavior. In the process, 'bad behavior' was not repeated. Those who had no social conscience, whom we now term psychopaths, committed horrific atrocities in that they were not psychologically troubled at all by the negative consequences of their actions on others—and hence why their cases were so shocking. Charles Manson, who killed an entire family, had relatively no cognition or empathy with regard to the suffering he caused.

Humanity's notions regarding the definition of 'core group' evolved over time, expanding in its ambit from family, to tribe, to nation, and the notion of the extension of rights to non-related others. Darwin sincerely hoped that some day this process would come to include other animals, in the consequent extension and expansion of well established human rights. The traditional Biblical arguments for morality were patently mistaken, and rights should be extended to other animals as well.

While both T. H. Huxley and Asa Gray noted uniqueness of the human species, with its faculties of language and cognitive abstraction, they also believed that human rights should also be extended to the rest of the animal world. John Dewey was certainly right he noted in the article "The Influence of Darwin upon Philosophy" (1910) that Darwinism would ultimately transform the treatment of morals, politics and religion—even while he did not specify exactly what these changes would be.

It goes without saying that the reaction by the Catholic Church to Darwinism was not positive. However, while Darwinism was generally disapproved of, the Church was relatively limited in its scope of action. This 'exchange' between the two worldviews has to be seen in the context of the Scientific Revolution, and particularly the infamous Galileo affair. The incident eventually became a 'shadow' on all consequent actions by the Catholic Church with regard to science.

It goes without saying that at the beginning of the early modern era, the scientific model had not yet been established, and had not widely diffused throughout Europe. The Catholic Church then claimed monopoly over 'truth', particularly the Jesuit order. The trial of Galileo was an example of an atrocious abuse of power that destroyed Galileo and his family. His daughter, the nun Maria Celeste was so anxious that she did not eat, leading to her death by dysentery—and was then buried in an unmarked grave by the Catholic Church. This could easily have also been Galileo's fate, which in turn would have killed the birth of physics at its conception.[147]

Fortunately, by the time the Darwinian theory arrived during the nineteenth century, modern science was already relatively well established, and was currently undergoing a process of professionalization throughout Europe. It was becoming institutionalized in universities, institutions, journals and was increasingly obtaining formal support by the state. There can be no doubt that the context in which these two worldviews interacted was radically different from that of its early modern counterpart.

One might even characterize the Catholic Church's reaction to Darwin as one of passive aggressiveness, wherein if a proDarwinian author lived and worked within its ambit, the Church could easily enforce any type of intellectual suppression by the sheer threat of expulsion, mischaracterization, or formal

---

[147] Fortunately, Galileo published two other works in Holland, still claiming the works were not by his hand for fear of even further retribution.

censorship through its use of the *Index of Prohibited Books*.[148] However, in those cases where the author occupied a social space outside of its control, then it would issue printed attacks in journals as the *Civilta Catholica*, that had a much more limited impact.[149]

In can be generally stated that, the further away an author lived from Hispanic countries, the safer he was from undue Catholic influence. This was particularly the case of those who lived in Protestant England where the Catholic Church had little to no weight and constituted just a tiny minority without much political power. There are a number of examples which closely fit within this framework.

George Jackson Mivart, member of the British Royal Society, wrote the first review of *Origin of Species*, and even personally knew Darwin. It is perhaps for this reason why Darwin was surprised when the negative review appeared. Jackson did not accept its principal notion of natural selection. In spite of this, Jackson's own book, *On the Genesis of Species*, claimed evolution was not incompatible with religion. In the essay "Happiness in Hell", he argued against the eternity of hell, for which he was specifically condemned by the Church for going against Catholic dogma. His book was placed in the *Index of Prohibited Books* in 1893, and, on instructions by the Vatican, was cruelly excommunicated by his own bishop a few months prior to Jackson's death in 1900.

John Augustine Zahm, a philosopher who had become Vice-President of the University of Notre Dame, wrote *Evolution and Dogma*, wherein he argued that there was no contradiction between Catholicism and Darwinism. Zahm called for a 'theistic evolution', wherein one should seek for proof of God not in miracles but in the law of nature. He defined evolution as an intermediary process, akin to the indirect hand of God—a position similar to that of Newton's in physics. However, this attempted resolution between the two world views was ultimately

---

[148] It is to be noted that the *Office of the Holy Roman Inquisition* had changed its name to the *Holy Office*, while retaining much of its medieval character.
[149] The journal was edited by the Jesuits.

incompatible with science, and perhaps for this was supported by Geremani Bonomelli, a bishop in Cremona. In spite of this, Bonomelli's own pamphlet was also placed in *Index of Prohibited Books*[150], and he was forced to recant. Zaham had also been supported by John Hedley, a Benedictine priest in Ireland, who in turn was criticized by *La Civilita Catholica*.

A third example is that of Dalmace Leroy, a French Dominican monk, wrote *The Evolution of Organic Species* in1887. While he recognized that Darwinism offended orthodoxy, he argued that the Church would ultimately have to come to accept it, as with the case of Galileo. Leroy, who would be proven correct over the long run, ended up being condemned by Catholic Church by an anonymous accuser, much like the Galileo case. Leroy was then forced to publicly recant his views in the newspaper *Le Monde*. His case is similar to that of the Italian parish priest Raffaeilo Cavieri of Quarante (near Florence). Cavieri was an enthusiast historian of science whom wrote the six volume *History of the Experimental Method in Italy* (1877) as well as *New Studies of Philosophy: Lectures to a Student*. As the prior cases, Cavieri argued that evolution could be harmonized with science, leading to his book's placement in the *Index* by 1878. He was also publicly condemned.

These cases suggest that the principal problem between the Catholic Church and Darwinism was that the Church used beliefs as markers of identity. Because belief systems change very slowly, this pattern of evaluation inevitably acquired a conservative reactionary character that was unwilling to reevaluate prior theological stances upon the emergence of new data. Inevitably perhaps, the Church's position became increasingly untenable as the scientific context in which its truth claims existed continued to change. It is also clear that the Catholic Church would have very much liked to persecute those who denied dogma, but increasingly could not do so as too much scientific evidence and public support continued to accumulate.

Yet the most tragic case is perhaps that of Tielhard de Chardin, a true scientist who was also a Jesuit priest. In spite of the fact

---

[150] Hereafter will be referred to as *Index*.

that de Chardin wrote one of the most sophisticated attempts to merge science and Catholicism, he was still was banned by the Catholic Church. His experiences best reflect the shameful intellectual history of the Catholic Church.

After bravely participating in World War I, for which he obtained the highest ranking in the Legion of Honor, de Chardin also became a geologist and a paleontologist. Studying in both England and France, his doctoral dissertation received an award by the French Geological Society. He then worked at the Museum of Natural History, the most important center of biology prior to Darwin, where renown biologists as Cuvier had worked. De Chardin was also a member of Academie des Sciences in France, and wrote some 200 scientific articles. While undertaking studies in China, he discovered *homo erectus pekinensis*, and later becoming involved in the minor fiasco of the "Piltdown Man", albeit as a minor player.[151]

De Chardin undertook the most serious attempt to reconcile Christianity and Darwinism, recognizing that Catholicism as a religious creed was becoming stagnant, unable to fulfill the needs its believers in the modern world.[152] He sought to 'upgrade' Catholic theology by its incorporation of evolutionary biology, resulting in *The Phenomenon of Man* (1927). He pointed out that man is the only creature aware of his own creation and with the capacity to actually mold his own biological development. De Chardin also noted that consciousness is the result of emerging complexity, reflecting an inverse relationship between simplicity and life, the latter which amounted to complexity of design. Greater complexity lead to unprecedented results which were greater than the sum total of their parts, borrowing from Jean Jacques Rousseau. As Lamarck, he believed that evolution was

---

[151] In the incident, bones were discovered which seemed to be 'missing link', but in fact was simply an orangutan jaw that had been filed down to fit a human head. Fluoride was used to used to identify the forgery. When de Chardin became famous in his later years, he was unjustifiably accused of the incident.

[152] He had not been the only thinker to attempt to reconcile religion to an ever increasing secular world view based on science, as noted by Hindu thinkers of the time.

teleological, in that it was headed to a particular point. In particular, he believed evolution also led to higher consciousness which would result in the formation of the *noosphere* or collectivity of minds, akin somewhat to the role played by the internet today. It goes without saying that de Chardin had a big influence in the new age movement of the 1960s, which formed part of computer revolution of the 1970s.

Influenced by Leibniz, de Chardin argued that people had to develop their own identity and individuation, akin to notion of cells in a body. Only when the component parts (i.e. cells) become specialized, does the biological body obtain functional coherence. Similarly, only when humans specialized, would humanity reach a new consciousness, a process that would culminate in the "Omega Point", which de Chardin associated with Christ.

In spite of the non-scientific and mystical nature of his views, there can be no doubt that his was first serious attempt to redefine Christianity in the context of science. De Chardin sought to create a new world view that, in contrast to its theological history, took into serious consideration the dynamic character of the universe. For de Chardin, this integration was desperately needed to breathe new life into the ideologically paralyzed Catholic Church, which had grown stagnant through its mere repetition of old and outdated ideas. While we will likely not agree with the specifics of his works, the significance and originality of the attempt has to be recognized. He was praised by many colleagues, including founder of the modern synthesis in genetics, Theodosius Dobzhansky. Julian Huxley himself was generally impressed with effort, even if he disagreed with its specifics.

Tragically perhaps, the work of de Chardin was ultimately rejected by the very two sides he sought to reconcile. A review by Peter Medawar was devastating, pointing to the overabundance of bunk science—a critique which would later be repeated by Richard Dawkins. As a Jesuit, de Chardin went to Rome to seek Catholic permission for the publication of his work, and was not only denied permission but was also forced to recant his views. Eventually Chardin ended up going to New

York in 1948, where his companion posthumously publishes his magnum opus.

What is tragic about his story is that the Darwinian interpretation of life was gradually accepted by the Catholic Church in a process which began in 1960. By 1996 Pope Paul II publicly recognized that evolution was not a theory, and in 2006 Pope Benedict XVI finally noted that so much evidence existed in its favor, that evolution should be accepted as a reality.

We should perhaps end the chapter by asking whether Darwin was an atheist. Did he seek to create an antireligious ideology? This is a complicated question to answer.

As a young man, Darwin believed in the traditional argument by design, and one of his favorite authors was William Paley, whom had written one of the most extensive treatments on the topic. Remember that Darwin's beloved university professor, John Henslow, was a natural theologian; it had been Henslow whom turn down the offer and eventually recommended Darwin for his life-changing trip to the Americas. In other words, Darwin was steeped in the religious tradition of his time, wherein nature and God harmonious coexisted together, and there is no reason to believe at the outset of his trip that these views would become as substantially modified as they did.

It is equally clear, however, that Darwin shifted his views over time, and was very hesitant to share this ideological change with his wife Emma. Prior to their marriage, he had even discussed the issue with his father, who counseled that she should not be told, noting the anecdote of a friend who lost his wife as result of it. Emma eventually did come to learn of Darwin's atheism, but hoped all of her life that she would also be able to change his opinion. Darwin certainly did not like arguing about the issue, and noted in a letter to his wife that he had shed many tears as a result. Darwin lived in a deeply religious society, and many of his loved ones simply did not share his views.

It is also important to note that his *Autobiography* was initially written only for his family, but was published a few years after by his son Francis. However, the book had been greatly edited, and all of the portions alluding to religion were removed from the original text. The original *Autobiography* did not emerge until

1958, published this time by his granddaughter, and it is here where the unquestionable atheism of Charles Darwin emerges.

It is also to be noted that his conclusions with regard to the conscience of animals was right on the mark; there is abundant evidence with regard to reciprocal altruism in primates. A 1964 experiment undertaken with rhesus monkeys at Northwestern University is surprising in this regard. The rhesus monkeys were taught that every time they pulled a chain, they would receive a morsel of food. At another stage of the experiment, they also learned that upon pulling the chain, they were electrocuting another rhesus monkey in a nearby cell. Such was the association between the chain and the pain caused to another conspecific, that all stopped pulling the chain and eating. Some went so far as to go without food for nearly two weeks rather than cause harm to another; this was particularly the case if the monkey pulling the chain had previously been its victim. The findings were consistent regardless of dominance hierarchy and irrespective of gender. Again, it is clear that the innate and 'deep' notion of human justice dates back over millions of years of evolutionary history.

It is certainly the case as well that conscience varies between species, in that they depend on a series of cranial subroutines to become activated, and as such are less well developed in the lower animals—and in this sense de Chardin was right with regard to varying levels of consciousness.[153]

There can be no doubt that the religious implications of Darwinism were enormous. The traditional view was defined by teleology, with a particular end, a designer, and wherein goodness and perfection prevailed. The Catholic Church's reaction to Darwinism had to be limited within the particular historical context in which the exchange emerged. The Church,

---

[153] It has been found that the notion that taste is specific to each animal species. Turkey vultures thrive in 12 day old carrion, as we do when we eat Thanksgiving dinner. That being said, they do reject it after this period, because the toxin levels have raised to too high a level. Snakes curiously hunt by eyesight, but will use their sense of smell to detect prey, even if physically wrapped around it. Rabbits trained to sense danger in one eye will not react if that danger appears in other eye.

nonetheless was still using dogma politically and took more than a century before finally accepting Darwinism. The Church's relatively mooted reaction, however, should not be taken to imply an acceptance of science, as shown by the case of Teilhard de Chardin, a scientist who undertook the most serious effort at the reconciliation of Catholicism and science, but whom was still rejected in his day.

The issue begs the question with regard to its reconciliation. Will the Catholic Church then see itself with the power to impose its beliefs on others, as Jesuits have historically done, pretending to be the ultimate guardians of truth? What happens when old truths become subject of debate, and their error is visibly demonstrated? The realistic interpretation is that the historical pattern of the Catholic Church's relation will likely repeat itself ad nausuem simply because the Catholic Church has too much financial and political power tied to its reputation. After all, it only survives on the good will of its followers, and any challenges to its reputation, are quickly and swiftly met like any other large and powerful institution—akin to that of the modern corporation.

Galen and Violence in the Post Classical World

As noted by Matt Ridley, the human mind is a difference engine able to quickly compare points. One result, however, is that conclusions and perceptions are not set within an absolute framework, but rather are established relative to their context. Positions of judgment are relative to its extremes, as in a Gaussian distribution curve of biological traits. For example, computer monitors on average used to be much smaller during the 1990s. A 15 inch monitor was considered the norm, but now appears small relative to the 21 inch plus monitor norm. This shift in baseline perception arises only because the context of judgment has shifted, specifically changes in the size of the largest monitors from 20" to 27". Today monitors that are 40 or more inches are also regularly available. The dynamic is referred to as 'framing' in the communications literature: how you contextualize something will have big impact on its initial impression and final valuation.[154]

This framing dynamic, where points of judgment shift relative to their context, also occurs at a much broader social level. Right wing Cubans have had long a relationship and integration with the CIA, specifically those who fled Cuba after

---

[154] To provide a personal example, I once participated in the organization of the centenary of my high school. A proposal had been made by another student for the official painting, which would be used as the cover for a brochure I was preparing. When the sagacious artist presented his samples at the meeting, he first brought out a rather horrific painting, which shocked and surprised those of us in the room. By contrast, the second painting, even if it was not that impressive, was so much better than the first that it appeared as a far more acceptable 'masterpiece'. It ended up being selected by the committee. The artist could now include it in his curriculum vita, which enhanced his social status and recognition of local achievement.

the rise of Fidel Castro. They have participated in a number of political assassinations, both at home and abroad. Some of the key individuals involved in the Watergate Scandal of President Richard Nixon, specifically those who actually broke into the offices of the Democratic party, were Cuban nationals. CIA affiliated Cubans were also key in the destabilization of Latin American governments after WWII, participating in the 1973 coup d'état of the Salvador Allende government in Chile. Allende then sought the nationalization of major US corporations, including the ITT corporation, and copper production, Chile's principal source of income, which served as the veins of the telecommunications conglomerate. Right wing Cubans demonstrate a violent streak at a group level with a low moral bar: the ends justify the means.

Empirical historical evidence thus shows that right wing Cubans have very different value structure from other Latin American nationals.[155] The unjustifiable extreme forms of violence visible in Cuban culture become more understandable when seen in their historical context. For example, slavery in Cuba was far more engrained than in Puerto Rico, which stood at some 40% of the population at its historical peak—around half of its Cuban counterpart.[156] As Franz Fanon noted, slavery is, by definition, a violent institution, of the dominant and dominated and of command and control relations. It is a far contrast to the relationship of equals seen in Classical Greek democracy.

We may also note that the Cuban wars of independence during the nineteenth century began much sooner than in Puerto Rico; the period was dominated by atrocious wars and civil conflicts. The Ten Years War between 1866 and 1876 devastated the countryside, while the war of liberation from Spain between 1895 and 1898 was characterized by horrendous cruelties. Gen. Valeriano Weyler, tasked with the pacification of Cuba, operated at an unbelievable level brutality from today's standpoint. During

---

[155] While historical violence undoubtedly existed in other regions as Puerto Rico, they did so at another far more reduced levels.

[156] It goes without saying that slavery was pervasive throughout the colonial Caribbean, showing much higher indices in other islands, as Haiti at the 90-95% range.

the twentieth century, dictatorships as those of Fulgencio Batista were no less brutal. As a US puppet, Batista's arbitrary rule led to a corrosion of public morality, which set the stage of the Castro revolution. Fidel Castro merely provided a spark to the enormous amount of unexpressed public resentment, visible from the fact that there were multiple points of revolt lacking a single central coordinating source.[157]

In short, Cuba's violent history produced a violent culture with a low moral bar. We should therefore expect a broader generalization of the phenomenon.

As societies become more violent, ethical standards should consequently decline due to the phenomena of framing. Such violence may emerge in the form of personal crime or more broadly at the social level in the form of warfare. Both are characterized by violence in the existence of physical or emotional harm, or injury done onto the human body or an individual's emotional state. Again, the reason for this cycle is that the contextual bar for ethical standards has been lowered, as the extremes have shifted. Even while the upper bar remains the same, the lower bar has been substantially reduced, either through an increase in the total quantity of routine violence or through the emergence of more grotesque forms of violence as those in civil war, so well depicted by Francisco Jose de Goya in his "*Desastres de la Guerra*" during the Napoleonic invasion of Spain (1810-1814).

The shifting ethical perspective can also be observed in US soldiers returning from wars in the Middle East as Afghanistan or Iraq. They typically suffer from post traumatic stress disorder (PTSD) as their wartime lived experience was very different from that of peaceful civilian life. Their perception of daily life had been drastically altered as a result of the war. They are naturally much more jittery, anxious and prone to overreaction—behaviors which are appropriate in a war zone but inappropriate when carried over into domestic life. The mean point of perception of their Gaussian curve had been shifted, and persisted in spite of returning back to the 'normal world'. As all madmen, it was as if

---

[157] It might be claimed that the Castro Revolution is a misnomer to a degree.

they were living in a different world—which cognitively they actually are. Subjected to much higher frequencies and intensities of violence, their standards of judgment have changed, showing higher incidences of both outward aggression in being more prone to be arrested, or in inward aggression given a tenfold increase in their likelihood of suicide.

Again, to restate our generalization, an increase in violence, should be typically followed by a consequent lowering of ethical standards at either the macro social or the micro personal level. How do we define ethics in this context? Ethics is typically framed within the norms of expected conduct and treatment towards others. The biblical golden rule of 'do onto others and you would have them do onto you' is a typical example of the notion of reciprocal altruism by Trivers. However, given that ethics is a branch of philosophy, changes in the social life of a community should also be reflected in philosophical systems, even if indirectly. In other words, changing social conditions lead to consequent modification of public norms and ethical criteria— a process that can be observed in the history of philosophy, in both Classical and Hellenistic periods, the latter which ranged from 323 to 31 BC.

The Classical period of Greek philosophy was profoundly influenced by the Peloponnesian War, around 430 BC. Socrates had served as a spear handler (hoplite) during the war, and could endure extreme pain. He often spent days in abstract thought, as if talking to the Muses, eventually forming a philosophy that was distinctly 'anti-empiricist'. Upon exile, his student Plato was influenced by Pythagorean notions. When Aristotle's wife's father is killed by the Persians, he flees to Lesbos, stimulating his biological studies. By the Hellenistic period, however, the amount and extremes of violence drastically shifted upwards with the increase in the number of military conflicts, as the Lamian War between Athens and Macedonia, or the Social War of 220 BC, where the hiring of mercenary armies became a regular practice. Cities sought to expand their territories and hence invaded each other continuously.

After Alexander's death in 322 BC the conquered territories were split between his generals. Antigone with his son Demetrius obtain control over Macedonia, which was less powerful, wealthy, or extensive than the rest.[158] There also followed an immediate period of political instability, that until settled was characterized by a great deal of violence and betrayal. This brief period can be seen as one akin to a hornets nest that has been disturbed, each attacking anything that moved. Leaders were elected, then killed, only for others to rise to power and be consequently killed in their turn, as if history had been sped up a thousand-fold. Changes that would take a hundred years occur in a very small time frame, and while its dynamics are too lengthy and complicated, we can provide a glimpse of its character.

All legitimate heirs of Alexander are killed. His son Arrhidaeus, who was retarded, is assassinated; Arrhidaeus himself had 300 men trampled by elephants for their discord. Alexander's surviving wife Roxanne and her second son are also killed. In this period, the victor was defined exclusively by power and military might. There was an increased use of the term 'king', which is extremely unusual in a Greek context. Yet the violence of the period is perhaps best typified by Demetrius, who eventually becomes the new king of Macedonia.

His father Antigone had depended a great deal on his son—a poor choice as it led to the loss of two-thirds of his father's empire. Demetrius implemented the largest movable siege tower, which had first been used by Alexander. These towers, pervasive in recent movies as "Lord of the Rings" or "Gates of Heaven," were key to attacking fortified cities. Demetrius called his own tower *Helepolis*: the taker of cities. It weighed 160 tons, was 7 stories high, had 8 large wheels, and required some 3,500 men to

---

[158] With his expansion to India, Alexander the Great extended Greek society and culture far beyond Athens. The character of Greek civilization also changes in a significant manner, with a larger urban scale, a more diversified population, and an influx of eastern and religious ideas. It has been argued that Greece was no longer itself by the time Rome took over; and out of respect for the Classical tradition it was not destroyed. The Hellenistic world was also characterized by a much greater amount of social and economic inequality. Whereas in Classical Greece all could serve in government, there is a consequent specialization in politics during the Hellenistic period. In contrast to the simple governmental structure of Athens, vast bureaucratic machines were created in Alexandria.

operate. There were multiple catapults at each level, with varying degrees of power: 180 lbs at low, to 30 lbs at high. Yet it turned out to be too big for its own good.[159]

In spite of his great losses, Demetrius was received like a king in Athens, as he did have a great naval victory to his name in the siege of Rhoades. The city had been built like an amphitheatre to the ocean. When Demetrius invaded with a huge number of ships, some 2,700 in total, the effect was psychologically powerful, instilling fear into the population. Athens began to seek freedom from Macedonia, and hires mercenaries for their causes, establishing in the process a horrific precedent which was noxious for the political life of the city state. Cities with the most money won military conflicts and grew politically powerful. This was to be contrasted with the traditional city state, whose citizen militia was limited by the city's total population. This social dynamic led to profound changes in Greek values, marked by a shift in the legitimacy of leadership to one based power rather than reason. The counterpoint to Demetrius is the case of Pericles, a key leader during the Classical period. He had been responsible for some of its most majestic glories as the Parthenon, an architectural masterpiece, and elevated the quality and tenor of Greek civic life.

By sharp contrast, Demetrius lived a life of debauchery in Athens. He is infamously known for his consorting with prostitutes at the Acropolis, and for his attempted rape of Damocles the fair. Demetrius even forced Athens to change her calendar for his initiation rites, which is perhaps the climax of personal arrogance. He was the epitome of the dictator, as shown by his absence of reason and his narcissistic focus. Demetrius also did not respect or follow tradition, and proceeded to kill most of the city's leading citizens—and is eventually in turn killed.

---

[159] In one attempted siege, the invaded city placed mud and water in the field of Helepolis' expected location. Consequently, the enormous siege tower became unmovable, ending up as an enormous waste of resources, time, and men. Big does not necessarily mean powerful.

The character of Hellenistic life was well captured by its historian Phillipus: "All conceivable calamities and continuous mutual slaughter have occurred in our time."

Violence could be seen throughout many aspects of Hellenistic life, and was generally regarded as an index of social power. The capacity to inflict pain helped to create and reinforce social hierarchies; under it, an emphasis on pecking order is taken to extremes, and might itself have been a byproduct of the increasing size of cities. Upon the existence of too many social relations, languages, and conflicting codes of ethics, it appears that violence simply became a quick and 'efficient' (but not effective) way of resolving social conflicts. Vestal virgins whom had violated sexual norms were buried alive under the city walls in Alexandria.

The violence of the period was also reflected in an epidemic of suicides in the island of Miletus, which had served as one of the origins of Greek philosophy. It appears as if the typical female suicide had grown in scale via network effects, and was creating a genuine social crisis. How does a social leader then place an end to such a pandemic if they could not possibly know the individuals considering suicide nor the particular personal reasons for doing so? It would have been unrealistic to talk to every woman in the island. The plague was curiously stopped by the use of reputation. When a victim's naked body was paraded around the agora or public plaza, the pandemic stopped immediately.

It is surprising to consider that an individual's consideration of future shame would overwhelm any temporary pain or discomfort. If the individual had been enduring so much actual pain in life, the imagined impact of any future humiliation would have been irrelevant and minor. In theory, if an individual was willing to take their life, they would have done so at any price. Yet by all historical accounts, the strategy worked. The use of reputation as tool of social control was in fact so effective, that it was prominently used throughout the Hellenistic period. Adulterous women were paraded seminaked on the agora. Military deserters sat in public for three days in women clothing, and were forbidden from holding public office or participating in politics.

In fact, the agora was the place where tyrants were most commonly killed. A public place is by definition an arena to reinforce public norms. Individuals tend towards their best behavior while in public, to positively influence their reputation in the cycle of reciprocal altruism previously described. In a public arena, all can see, and become aware of each other's behavior.[160] The assassination of tyrants in the agora thus served as a form of political legitimization. The audience becomes complicit in the crime; if they do not object immediately, it is then hard to complain later. As the assassination occurs before the public eye, it is given moral status by its implicit sense of public justice; successful assassins were hailed as heroes.[161] (There is obviously also the factor of fear, as anyone who opposed the assassination could also be killed on spot.) These assassinations actually received their own term: *kikaios phonos* or 'just murders', and statutes erected in public places, as in cases of Hermodius and Aristogeiton. When asked for the best type of bronze, their statues would be used as exemplars. There is the case of Timoleon who killed his brother, the tyrant Timophanes. As the latter was often surrounded by bodyguards, he had been only person who could get close to him. The purpose of such assassinations was simply to remove those who blatantly harmed the public good; to remove those who threatened the freedom and wellbeing of the community.[162]

The violence of the Hellenistic period is reflected in its art, poetry and architecture. While it had existed since classical

---

[160] Today, the modern state uses violence via a police force to reinforce public codes.

[161] Inversely, assassinations of public officials not committed in public were regarded as an act of treasons, for which its perpetrators were often brutally killed.

[162] Incidentally, if the assassination did not occur in public, it was not recognized as meritorious, as occurred in the dawn killing of a Roman official by Damon, who himself was then killed at a bathhouse (a public place). As the assassination had occurred at the early morning hours of the day, when nobody was present, it thus lacked it public moral aspect.

Greece, its cultural presence was limited as it was seen as inappropriate to glorify oneself. However, there is a gradual shift in tone, an artistic change initiated by Lysippus, who was court sculptor for Alexander. His depiction of the Battle of Granicus in 344 BC captures this spirit.[163] In the increasing portrayal of hunts and actual battles, there is an increasing depiction of the horrors of war. As in Goya's artwork, the faces revealed that war was not a pleasing event, a trend which is epitomized in the Gigantomachy frieze at the Altar of Zeus in the Pergammon.

The increase of violence ultimately had an impact on Hellenistic philosophies. A look at the Greek philosophical schools is illustrative.

With the fall of Alexander, control of Aristotle's Lyceum was given to his good friend Theophrastus. To repeat, Athens sought to remove itself from the Macedonian yoke, and consequently passed laws requiring the prior approval of all educational leaders by its legislature. The Athenian government actually dismisses Theophrastus from the famous school, but he had been so popular that the entire school was emptied of students. The surrounding shopkeepers complained that the loss had been followed by a precipitous decline in commerce. The government was eventually forced to reinstate Theophrastus. With his return, there arrived a new philosopher: Epicurus, now regarded as the founder of Epicureanism.

While Plato's Academy still continued to operate, a broad general decline in philosophy could be observed. Greek philosophy used to be dominated during the Classical period by three principal areas, metaphysics, natural philosophy (science), and ethics, but was reduced only to the third element (ethics)— and then somewhat in a questionable state. The leaders following of the Academy tended to be nihilists, somewhat ironic when one considers they were intellectual heirs to one of the greatest philosophers that has ever lived (Plato). Pyrrho distrusted senses and reason, pointing out that any argument could be put forth to support any position. This view was taken to extremes, as in the

---

[163] Nearing the defeat of Persians, Alexander charges ahead of time using his cavalry. He had been expected to get stuck by a river, but instead crosses an undefended pass, taking the enemy by surprise.

case of Carneades, who in 169 headed the "New Academy". He was sent to Rome, but remained there for two days. The first day he argued for a particular position, only to argue for the directly opposition position the following day. Cato immediately orders his return back to Athens as he saw such nihilism to be a grave danger to public morals.

Cato had a very valid observation. Nihilistic philosophies naturally led to moral nihilism as well. If nothing can be known, then no viable ethical stance can be taken, as any or all stances are equally viable. Under its rubric, morality and immorality merge, in that no distinction between them can be philosophically made. Cato's reaction was perhaps that of a practical man who detested creation of futile paradoxes, which more often than not confuse rather than clarify.

The nihilistic philosophers do illustrate an important point: belief in the existence of truth is a precondition for the existence of ethics. Their counterexamples in the Classical philosophers (Socrates, Plato, and Aristotle) are equally illustrative. Their belief in the existence of absolute truth afforded their philosophies strong ethical stances. In this, Socrates is perhaps best example. He believed in world of Ideal forms, in a world of absolute truths so typified in mathematics, and ended up becoming principally a philosopher of ethics rather than one of natural philosophy. Socrates continually asked how is one to live the good life.

Epicurus's position is perhaps not as extremely anti-ethical as is often presumed. While epicurean philosophy is usually depicted promoting one's removal from society, this view is somewhat exaggerated. While he did personally have no interest for politics, he was not averse to using public institutions, as the public records office in the Metron where his writings and documents were guarded for posterity. It was a 'negative' philosophy in that it sought absence of pain through *ataraxia*, or 'tranquility'. While Epicurus did recognize that society was necessary, in that society built systems that brought peace through its military and finances, he also sought to limit its detrimental impact, specifically that caused by social inequality which resulted in greed and envy. Since gods were defined as entities free from harm, Epicurus sought to obtain the same internal security in his philosophy. To speak as a god was to speak steadfastly, calmly, and serenely—absent of anxiety.

Epicurus believed that philosophy made men more divineand that the divine existed in all persons. If all men studied philosophy, it would create a peaceful world by the elimination of gross error. The greatest of all goods for him was peace—a stance which is an obvious reflection of its violent social setting. In this he also inverted the traditional Classical notion of the *polis*, by focusing his attention on friendship in the creation of ethics and equality in the establishment of happiness. While Aristotle claimed that only equals could be friends, as social differences would ultimately be abused, Epicurus believed that friendship helped to reduce social inequality, which to him was a principal cause of violence. He pointed out that most hate that which they perceive as different: different economic classes, ethnic groups, and genders. Friendship helped an individual to focus away from differences with other individuals by emphasizing their commonality. When dying from kidney stone, it was his remembrance of conversations with old friends that helped him cope with the enormous pain he was suffering.

While these ideas might be seen as abstract irrelevancies by some, it is to be noted that Epicurean philosophy is far more important than appears at first glance. A precondition for the existence of any politics, so necessary to modern life, is the rational deliberation of the subjective viewpoints between free individuals. Violence by definition 'kills' political systems as it squashes the open expression of opinion and the possibility of compromised mutual accords. Violence hence normalizes the inequalities of power, the best example of which are the physical abuse of slaves, which make it unrealistic to expect genuinely humane interaction between the parties involved.

Another important Hellenistic philosopher was Zeno, founder of Stoicism. As Epicureanism, stoicism was an atomic philosophy. Zeno himself was humble to the point that he could be seen as the Franciscan monks of period. He abided by a vow of poverty, slept wherever he could, and wore ragged clothes. These features are striking as Zeno had been a wealthy merchant, saving up to 1,000 thalers, before migrating to Athens. When his ship crashed, he lost his entire fortune. He made his way to an Athenian bookstore, where he read Socrates and became enthralled by philosophy. His school was named after the *Stoa Poecile*, or pointed porch where he would talk with anyone who desired to do so. Incidentally, he disliked speaking to youth, pointing out that since humans have only two ears and one mouth, they should listen more and speak less. He was nonetheless prized by Athens for teaching the temperance, paying for his funeral and honors upon his death.

One curious feature of stoicism was its distinct cosmopolitan outlook, which as in the case of Epicureanism, had been greatly influenced by its social setting. It was a philosophy made for a newly emerging urban world, and would served as a critical backdrop for the formation of Roman law. The religious aspects seen in Epicureanism were expanded in Stoicism, and widely adopted by Jewish rabbis. Many key passages from the Bible are near exact copies of stoic texts. The references to God parallel those of Zeus.[164]

---

[164] *Thou, O Zeus, art praised above all gods;*
    many are thy names and thine is all power for ever
The beginning of the world was from thee:
    and with law thou rulest over all things
Unto thee may all flesh speak:
    for we are thy offspring
Therefore will I raise a hymn unto thee:
    and will ever sing of thy power
The whole order of the heavens obeyeth thy word:
    as it moveth around the earth:
With little and great lights mixed together:
    how great are thou, King above all for ever!
Nor is anything done upon the earth apart from thee:
    nor in they firmament, nor in the seas:
Save that which the wicked do: by their own folly.
But thine is the skill to set even the crooked straight:
    what is without fashion is fashioned and the alien akin before thee.
Thus hast thou futted together all things in one: the good with the evil:
    That they word should be one in all things: abiding for ever.
Let folly be dispersed from our souls:
    that we may repay thee the honor wherewith thou hast honored used:

Both philosophies sought to help individuals navigate the new urban social spaces created during the Hellenistic period. Epicureanism sought to reduce violent behaviors via both friendship and philosophy, and is a reflection that the actual requirements for human happiness are few: food, shelter, and friendship. Today it is generally accepted that the obtention of wealth beyond a certain point renders no net contribution to happiness and well-being, which primarily stems out of relationships with those around us—particularly in the reinforcement of reciprocal altruism. Human social bonds are those which give life its deepest sense of joy. Stoicism, for its part sought to increase mutual collaboration via cosmopolitanism. By integrating many eastern elements it eventually became the philosophical basis for an emergent Christianity. In their principal features, both had been clearly influenced by the prevailing social atmosphere out of which they emerged.

Singing praise of they works for ever:
*   as becometh the sons of men.*
-Cleanthes (on Zeus)

Naturalists and Colonial Violence

The number of naturalist explorations that were undertaken in the Americas is impressive. There were periods of concentrated exploration, as that towards the end of the eighteenth century. While many prior missions had been proposed, they were now granted royal sanction and approval. Some of the more important included those by Mutis in Colombia, Ruiz y Pavon in Peru, Malaspina in the Pacific, as well as Sesse y Mociño who touched on Puerto Rico shores. From 1760 to 1808, there were a total of 57 expeditions, a far higher rate than had been previously seen, which in turn meant more information of the Americas that was returning to Europe. Ruiz and Pavon alone sent 2,650 illustrations between 1779 and 1788.

These had been preceded by an earlier wave at the beginning of the Colonial period in the sixteenth century. Some of its explorers included Cristobal Colón, Michele de Cuneo, a friend of Colon, and Melgarejo[165], whose report on the island of Puerto Rico reads like a modern day travel guide. Some of the most important had been the journeys ventured by Bartolome de Las Casas and Gonzalo Fernandez de Oviedo. Oviedo's first work the *Summario*, published in 1536, followed by the first volume of his *Historia Natural* of 1535. Both Las Casas and Oviedo had an avid disputes pertaining to the character and nature of America's indigenous population, which led Las Casas to effectively block Oviedo's later publications.[166]

---

[165] Juan Ponce de Leon

[166] Other travelers included Francisco Hernandez who traveled to Mexico between 1571 and 1577, perhaps the earliest work that was the most comprehensive. Tragically, his materials were lost to history during a fire at El Escorial in 1671. One of the most interesting travelers was Jose Acosta, a Jesuit father who traveled to Peru and Mexico in the latter half of the sixteenth century. Acosta sought a reconciliation between Europe's traditional Greco-Christian learning and the perplexing evidence from Americas.

What is perhaps most surprising is that in spite of the fact that hundreds, if not thousands, of naturalists had visited the Americas and other parts of the world throughout the entire Colonial period, none of them came up with notion of natural selection.

Facts do not in and of themselves reveal theory or 'truth'. To see with the eyes is not necessarily the same thing as 'seeing with the brain ', as Al-Hazen noted. We tend to see only that which we already know, with mind full of preconceptions. For this reason, the demonstration of 'facts' does not in and of itself constitute a necessary 'proof' or leads to a change of belief in the observer. Evidence, psychological speaking, is not irrefutable proof of hypothesis, a notion which could just as well be applied to the Scientific Revolution as well.

Therefore, before the science of biology emerged as a discipline, many other factors needed to occur; the discovery of new species was a necessary but insufficient cause to its existence. The diversity of specimens needed to be amplified, which in turn were obtained by the many voyages to the Americas. An appropriate classification system needed to be created, culminating in the successful work of Linnaeus with regard to plant species. A similar system also needed to be established for fauna, which was achieved by Georges Cuvier, using his notion of subordination of parts. In other words, the Darwinian evolution could only emerge until many other steps had been previously achieved.

Remarkably, at each stage in the development of the science, biologists and protobiologists were presented with their own unique 'bioethical' issues. Animal experimentation was not yet a factor in natural history, even if violence was a common experience during the Colonial period. As previously seen in the case of Hellenism, different levels of violence led to shifting ethical standards, which were as a rule not particularly rigorous.

During the first encounters, violence was a necessary feature of daily life. It might perhaps be useful to remind the reader that there were no supermarkets at every 'street corner' where one could readily obtain food. There were no *Pueblos* (Puerto Rico), *HEB*s (Texas) or *Krogers* (Ohio), and much less a restaurant at every corner. "Fast food" might meant fruits and berries or a month-old maggot-infested piece of bread. Hunting and fishing were the routine requirements of daily sustenance.

Oviedo, known for having a good appetite, examined the many creatures he ate. There is the funny story of his encounter with the pitahaya; he incorrectly believed his red urine implied poisoning and impending death. Darwin was perhaps no different. The rhea birds eaten on the trip were analyzed to identifying varying species in nearby territories.

Violence was also a regular part of European interaction with local Native-American inhabitants. The ontological and ethical status of the Indian was unknown; was it human or a kind of ape? Did it have reason? This was the source of the dispute between Las Casas, a prior colonizer in Santo Domingo, and Oviedo. Both as young men had seen Columbus arrive from his first trip, and as young boys had both been awed by his discoveries. Oviedo had then been part of the royal Spanish entourage, serving as companion to young Principe Juan, whom died under mysterious circumstances soon thereafter.

The ontological/ethical status of the native American had also been studied by Acosta, who recognized a relatively high level of civilization in spite of its differences from European civilization. The complexity of their cultural objects and social structures suggested that they had enough 'reason' to accept faith, and hence to benefit from European legal codes upon their conversion to Christianity. If they, however, had no rationality or reason, they would have not been able to convert to Christianity, and receive the social blessings which proceeded therefrom.

Yet while the academic-theological debates on the nature of the native American were going on in Europe, abuses readily occurred on the other side of the Atlantic. Cuneo, whom had joined Columbus in second trip, regularly mentions the atrocities committed by the Spaniards in the most nonchalant manner, as when an Indian's head was cut off. The extreme forms of violence in Cuneo's description uneasily existed side by side with impartial depictions of the flora and fauna of the region—a shocking irregularity when seen form today's lenses. Cuneo lived in a different moral realm than our contemporary one.

Violence also became institutionalized in the Spanish crown, well illustrated in the case of Christopher Columbus. One might think Columbus would have been lauded as a hero decades after his first return. There mere act of crossing the gates of Gibraltar was a tremendous act of valor, as few ever returned when crossing from the Mediterranean to the Atlantic. When he first proposed the trip, many thought Columbus was crazy, and he was abundantly attacked for his proposals. Yet, in spite of this sacrifice, when the Spanish Crown became aware of the extent of his discovery, the avarice was of such a degree that he was abused and cast aside.

In his *Letter from Jamaica, July 7, 1503*, we learn that he had been promised governorship over an enormous area of his discoveries, ranging from the North to the South pole. This geopolitical incentive hence gave Columbus's voyages their distinctive trait of attempting to discovery as many places as possible; the more places seen, the more he could claim for himself and his family. His was principally a costal journey that typically did not venture far inward. His stereotypical portrait is always characterized by the beach, interacting with the Indian. However, upon becoming aware of the sheer scale of the discovery, the Crown reneged on its agreement. Colombus's letter notes that although he had placed lives of his loved ones at risk, particularly his sons and friends, they were jailed and grossly mistreated. Those who had initially criticized him were all too quick to take advantage of the discovery, claiming lands to quickly get rich. Columbus, worn down from his trips to the Americas, dies relatively soon after arriving from last trip, at the early age of 55.[167]

---

[167] Today Columbus is routinely attacked as an "*usurpador*" by historians who over-idealize the alleged existence of an indigenous pastoral lifestyle, which never existed. Colon was just as much victim of the monarchy as the very indigenous he sought to protect, and in this he was a precursor of a much larger trend. Most naturalists typically suffered some type of abuse from the Spanish crown, and is perhaps what makes their effort so tragic. Their journeys were perilous ones, full of great sacrifices, and upon their return had to deal with royal and Spanish avarice.

The uncertainties of the weather as hurricanes were a regular feature of their travels. One storm blew Columbus's ship off course. There were the uncertainties of the tribes they encountered. Caribes routinely attacked, and settlements as Navidad did not survive. There were geopolitical uncertainties as well. Naturalists had to send materials gathered back, but their packages were often subject to attack and confiscation by pirates or rival European powers. Their manuscripts might be stolen, only to be republished under another author's name. The work of a lifetime could be voided in instant. The voyages themselves took up large portion of time, as years and decades could be spent on a trip, much as Homer's *Odysseus*. Their adventures were no weekend affairs, and a single trip from Europe to the Americas might take 13 months of seafaring. There were countless other injustices suffered by conscientious naturalists. Yet, in spite all of this, the Spanish crown routinely reneged on its own promises and contracts—and they weren't the only ones.

Formed during Counter Reformation in the Council of Trent 1593, the Jesuit body was given the function of protecting the faith against Protestantism, and developed a value system in which learning was a way to God. In the Americas, the Jesuits interacted with countless indigenous tribes, who routinely fought each other and might not necessarily speak the same language. Each of these tribes held their own distinct body of medical knowledge accumulated over centuries regarding the medicinal value of particular plants. If something worked, it was passed down over generations, in a cultural development akin to that described of capuchin monkeys in the *Planet Earth* documentary. Once acquired, the cultural value of such medical knowledge could be easily taken for granted.

The distinct ethical dilemma faced by the Jesuits was the following. They wanted to obtain medical knowledge while at the same time convert as many Indians as they could to Catholicism. The strategies of both goals, however, were mutually contradictory. The Jesuits opted to betray their religious chargees, and not only published the findings as their own, but distributed these via an institutional network which was then used to co-opt indigenous communities on a continental wide basis—a procedure known as cultural genocide. In other words, the Jesuits used the freely given indigenous medicine to elevate their own personal standing and convert the very same tribes whom had freely shared this knowledge to Catholicism. Had the indigenous tribes collaborated amongst one another, the Jesuits would have never been able to capitalize from indigenous medical discoveries, as these would have already been readily distributed between the different tribes.

The Jesuits had two distinct advantages over their chargees. Indigenous tribes did not have a written language nor did they have organized institutions of schools and churches coordinated into a continental wide network. At their peak in 1750, the Jesuits had 800 *colegios* (schools) and missions. These gained an advantage simply out of the sheer scale of their organization. Each Jesuit mission could obtain a specific remedy of the indigenous medical men, which was then written in a book or pamphlet and disseminated throughout their entire collective structure. By contrast to the diverse warring tribes, the Jesuits were able to accumulate knowledge into a coherent body, which had originally been retained only within specific Indian groups to which it belonged. This body of knowledge would then be used for Catholic indoctrination in that the Jesuits feigned to be new medical men of sorts, and in effect betraying the very individuals who had helped them obtain knowledge in the first place. It is perhaps no wonder that so many Jesuits were brutally murdered in the jungles of South America; plagiarism and cultural genocide are not issues to be taken lightly.[168]

The colonial appropriation of indigenous medical knowledge had enormous geopolitical implications. In other times and places, this knowledge allowed imperialism to emerge, particularly during the nineteenth century partition of Africa. Prior to its worldwide diffusion, tropical disease served as natural barrier to the interior of the continent, and it was not until the nineteenth century that a gorilla was seen for the first time. While

---

[168] These dynamics are akin to those between the CIA and Latin America, where advanced technology provide an unequal advantage in the relationship between the two parties. This advantage is wonderfully depicted in the movie "John Carter" (2012).

cinchona saved many lives from malaria, it was distinctly used for geopolitical ends, detailed in the works of Daniel Headrick.

These cases of colonial power resemble female rape and male cuckoldry, because of the removal of agency from the historical actors. To remind the reader, the female does not select mating partner and when the male is cuckolded, his resources go to raising another man's children. In this particular case the Jesuit are acting like the CIA in that they are using their greater organization and more advanced technology to alter the historical destiny of groups to its own particular ends.[169]

It is a tragic irony that one of the earliest and most 'scientifically' trained naturalists, Francisco Hernandez, became one of the most mistreated by the Spanish crown. Hernandez had studied medicine in leading Spanish medical institutions, becoming physician to the king. His life's work of 1577 would not appear until 1790, more than 200 years after its creation. In 1611, Galileo had seen the collection in the hands of his colleague, Federico Cesi, and had been very much impressed. More than 3,000 new species had been discovered by Hernandez, including the papaya, the avocado and the peanut. In his drawings, he had followed a rigor akin to that of Leonhart Fuchs (1543), one of the founders of European botany. Such a high level quality would not be seen until the work of Sesse, some 350 years later. While today we take the faithful representation of

---

[169] It goes without saying that had the Jesuits not been there, tribal medical would have taken different historical route. In competition with each groups, themes might have lead to success of a single overarching body, akin perhaps to the Maya codes of astronomical knowledge. The Jesuits violated the trust placed on them, in not allowing indigenous groups their own autochthonous development. Fair competition this was not, lacking equality of conditions. However, we should not necessarily idealize indigenous tribes either; their lives were one of extreme violence. The division between the groups and their constant animosity might have simply led to the historical loss of such knowledge had the Jesuits not taken the steps they did.

It might be pointed out that biological classification had yet to be invented, and pre-Linnaean botany was characterized by the same mix-ups and confusions seen in chemistry during the alchemical era. So much false information and speculation existed, that information tended to be rejected by the Royal Society.

nature for granted, given the widely available cell phone photography, prior to this, all faithful representation had to be done by hand, hence making artists an essential auxiliary aspect of Colonial scientific activity. As soon as Hernandez returned from Mexico, he was betrayed by the Crown—losing control over all of the materials he had so diligently gathered in the Americas.

The same might be said of Iñigo Abbad y Lasierra. A Franciscan monk influenced by Enlightenment authors, he traveled to Puerto Rico in the decade of 1770. While his original aim had been to write dictionary of the Americas, as most books on the region were written with military and commercial point of view, he shifted his focus. A 'philosophically oriented' book needed to be created. His *Historia Geográfica, Civil, y Natural de la Isla de San Juan Bautista de Puerto Rico* (1782) details both its historical origins as well as the abundant flora and fauna of the island.

Yet Abbad y Lasierra ran into problems with the law. He was accused by the military governor of owning a slave and of partaking in usury by charging 5% interest on a local loan. Certainly Abbad's treatise did not speak well of the local population—and might itself account for the poor treatment received in the island. Abbad y Lasierra observed that Puerto Ricans did not take full advantage of the of plentiful resources around them—a viewpoint that would be adopted by US colonial governors during the twentieth century. Puerto Ricans were poor amidst a land of plenty, and traced the source of their poverty to their own lack of effort and education. Puerto Rico had gold, but local mines were only superficial and were worked without intelligence (specifically the science of mineralogy). In contrast to the cultivation of pepper in Jamaica, in Puerto Rico only culled rotten fallen trees, and hence its relatively low production. For Abbad y Lasierra poverty was a product of culture.[170]

---

[170] A few quotes helps provide the general tenor of Abbad y Lasierra's work: *"No hay que admirar la lentitud con que estos isleños adelantan sus conocimiento en esta ciencia (botánica)." "Con la misma indiferencia [los Puertorriqueños] miran las producciones de otros árboles, maderas útiles y resinas. ." "En esta isla hay pocos naturalistas que conozcan la virtudes de*

However, there can be no doubt with regard to the cultural impact of the colonial system under which Puerto Ricans lived. Positions of power were only held by *'peninsulres'* from Spain; the law tended to be imposed from above by Spain. The best critique of the system is perhaps that provided by Alejandro Tapia y Rivera. The policy of *'baile, botella y baraja'* denied the creation of higher educational institutions in the island, with tremendous implications for local sovereignty, which the Spanish monarchy feared. An ignorant people are always easily ruled; the absence of universities meant that local interpretations of their own reality count not be created. Those who did have the funds to afford an European education, suffered a double imposition. The forced travel to Spain typically led to a type of indoctrination; inversely, good candidates as Roman Baldorioty de Castro and José Julian Acosta were often tempted to remain in Spain, thus further thwarting the local investments in their education.

This chapter, as the prior one, has sought to provide some of the historical underpinnings in which ethical decisions reside. They do not occur in some metaphysical vacuum void of any historical context, but are rather relativistically set—as sociobiology would indicate. While it would be ideal to proclaim a set of universal historical principles, to then be established for the future of humanity, such claims would be grossly negligent and useless from a standpoint of social policy. While philosophy is a useful tool in the analysis of ethical points, it is useless from the point of view of policy—particularly when applied to biology proper.

The history of biology is so much broader than just laboratory work or animal experimentation, and much of it had to do with the institutional contexts in which they were situated: the logistics of exploration, discovery, and publication. The Spanish crown's policies were erratic and arbitrary, in that they generally did not recognize value of science until fairly late in the Colonial period. These social factors also profoundly impacted the

---

los vegetales, ni físico que sepan distinguir sus virtudes y accidentes a que deben aplicarse."

reception of the knowledge of Hispanic exploration into the development of biology in places as Germany or England. One of the most famous 'naturalist' was the Jesuit Athanasius Kircher, who espoused all sorts of fabulous claims, including that of evolution. His arguments, however, as Acosta's were not ultimately founded on empirical evidence, but were profoundly shaped by religious presumptions. Myths also abounded; large anacondas grew from their skulls.

The number of animals certainly increased with the greater exploration of the Americas, and placed much pressure to the formation of a coherent classification system. Aristotle had identified 500 animals, Theophrastus 500, Discorides 600 (*De materia medica*), and there were some 1,600 species by the end of the Greek period. In 1623 Caspar Bauhin had catalogued 4,000 plants, but by 1700, tens of thousands of species had been accumulated. John Ray's three volume work had 18,600 plant species; by the time of Linnaeus in 1750, more than 70,000 living organisms had been identified. There can be no doubt that the exponential increase put pressure on the extant classificatory system, as 500 is perhaps the largest number of items manageable by the human mind. Many systems contained a genera of 500 items.

The reason others aside from Darwin did not discover 'natural selection' should be fairly obvious by now, as the case of Abbad y Lasierra of 1782 clearly shows. The totality of animal forms were enormous; if botanists could not find a coherent structure, neither would a Franciscan monk were he to try, which he attempted to a degree. His general emphasis on utilitarian considerations, however, shifted his focus away from the issue. Common philosophical presumption such as a static universe or the notion of plenitude would have similarly inclined any thinker away from the effort. Given that all possible animal shapes existed, any variation in form did not necessarily create an anomaly within its paradigm. New forms were just more shapes of the universe, and did not appear as facts that needed explanation.

This presumption extended into ethical realm, in that unethical actions towards the creatures of the New World, human or not, did not represent a quandary to be resolved. Moral individuals as Bartolome de las Casas were exceptions to the rule of colonial relations.

Part III: On the Abuses of Biology

Prisoners and the information game

In the context of evolution, it is at first difficult to account for the origins of modern society, which by definition constitutes a large social groupings of non-related individuals. While it the most common setting of human communities, most species in nature only collaborate between kin of closely related individuals. Hives of bees, ants, and other insect are typically composed of closely related sisters. No human community could ever be as cohesive, and most attempts would fracture. Hence the commonality of human societies is a puzzling one; why have they not fallen apart?

The Germans point to a similar issue in what they call "Adam Smith's paradox". On the one hand, in the *Wealth of Nations*, Smith shows how individuals are utterly driven by self interest, which produces a coherent whole out of an aggregate mass of individuals. In each seeking their own self interest, a macrosocial 'invisible hand' is created, which guides the collective.[171] On the other hand, in the *Theory of Moral Sentiments*, Smith pointed to the goodness of humanity. Individuals are commonly driven by goodness, as the love of family. Smith's notion of "beneficence" was rather ironically adopted by Herbert Spencer. The two stances thus constitute a paradox for German thinkers, who noted that Smith proposed humanity was driven by both good and evil, by self interest on the one hand and by altruism on the other.

It is clear that the "long shadow of the future" (Robert Axelrod) plays a role in these dynamics. We are concerned about how others think about us, and cannot separate ourselves from the society of which we are a part. For much of human history,

---

[171] The baker cooks because he wants other goods; as he is most efficient in his baking, he can produce more baked goods than other at a lowers cost, which benefits the community overall, and in turn increases the diversity of exchanges.

Rodrigo Fernós

humans lived in tribal groups, the rejection of which would entail death. Similarly, single male chimpanzees have a high propensity of becoming victims to murder, if found within a neighboring enemy's quarters. We depend very much on others for our well being, to which we turn in time of need and help, and in this social relations as friendship go far beyond money or status.

Friends can become allies that last a lifetime, and are built in emotions which are positive for one's long-term well-being and survival. Albert Einstein is perhaps illustrative. Today Einstein is an icon, but Albert Einstein had not always been the world renown 'Einstein'—had it not been for his good friend. Marcel Grossman saved him on three important occasions. Grossman provided detailed class notes so that he would not fail the university exam, and also saves him from financial ruin by getting him a job as patent officer. Finally, Grossman establishes the mathematical foundation on which general relativity is built. While Grossman knew the mathematics of Gauss and Riemann necessary for the task, Einstein did not. There is no doubt that without Grossman's help, Einstein could have ended up a street bum.

Our reputation is important to this social process, but its construction rests on information exchanges which are obtained through long interaction with others. As Matt Ridley points out, the brain is a repository of "whom owes what", and in turn determinant on whom is a 'friend' and who is not, as the case of tropical vampire bats. Not all bats successfully hunt by licking ruminant blood, particularly young ones with little experience. If these bats go too long without eating, they can easily starve, and hence often receive 'gifts' by more experienced bats. However, the community is thus subject to the problem of the typical *free rider:* individuals who obtain food for free without attempting to even fend for themselves, common in all societies. For vampire bats, this problem is easily resolved by a simple ritual: the licking of the stomach, which constitutes an information exchange and claim verification.[172]

---

[172] By observing how distended a bat's stomach is, another can accurately measure the validity its calls for food.

180

Information exchanges and the verification of data are tremendously important in all social relations, human and non human alike. People are 'ethical' and 'good' to a great degree because of information, or a record of how they have behaved in the past, which in turn affects the reaction of others in a society.[173] There are many examples which illustrate the role of information on ethical behavior. When asked to share $100 with a stranger, the amount is correlated to the presence or absence of information; a consequent decline in information lead to an exponential fall in sharing. Rather than the typical equal ratio distribution of free money (50%), only a minute fraction is shared (1%); in the absence of information, the reputation for stinginess does not propagate to the rest of the community under these conditions, and hence an individual's positive reputation is not detrimentally affected.

We can make a broader generalization from this. Given that prisons are typically social settings wherein information is absent, the propensity for abuse is exponentially higher than in the free community.

Jeremy Bentham believed that best prison is that which contained a 'central eye' about jails cells, or a 'panopticon'. Given that they were under constant surveillance by an entity they could not observe (and hence detected when they were not being observed), all prisoners of the panpticon would be inclined to be good all of the time. However, sociobiological theory clearly shows the flaws of Bentham's reasoning. Typically, very little information reaches the outside of a prison, which in turn encourages the bad behavior of its participants, officers and prisoners alike, by definition. All sorts of atrocities as a result characterize the prison system, and it is in this context where worst breaches of medical ethics can be found.

Under these conditions, the best prison is that which does not exist in the first place.[174] Prison are relatively new in world history, and there are many doubts as to its benefits. Historically,

---

[173] If individuals have a reputation for goodness, they more likely to receive help in the future.

[174] If it exists, it should be as open and visible to the public as are the prisoners in the panotpicon.

'criminal' individuals were exiled or removed from their communities of residence, as in Classical Greek society. The advantage of this tactic is that public information was still present. In fact, some of the worst breaches of medical ethics have occurred in prisons, typical cases of which are Nazi concentration camps. These were actually industrial centers. It was said that their prisoners had been placed in the Budan diet, which squeezed the life out of a prison, as if compressing a metal ore.

The absence of information drastically encourages the abuse of power. Even a tyrant will want to be perceived as good to others, as it is a key in establishing his legitimacy, which is ultimately the source of his power. Governments who have systematically committed 'evil deeds are well aware of geopolitical impacts of such actions, and actively seek to prevent awareness or diffusion of this information when it occurs. All of these human dynamics represent clear evidence of the 'long shadow of the future', and its powerful impact on human behavior and human ethics.

Two cases will be discussed in this chapter. The first is that of Unit 731 run by the Japanese government in Manchuria during WWII. The crimes committed were so horrific, that anyone who ventures to study them must be psychologically prepared for its images of human vivisection, such as the case of a European pregnant woman. If every major war of the twentieth century has had a particular science attached to it, WWI known as the chemists' war and WWII and the physicists' war, we could suggest that World War III, if it occurs, will end up becoming the biologists' war.

The second case we will look at is that of the Guatemala syphilis study by the US Public Health Services right after WWII (1946-48) where countless victims were infected with sexually transmitted diseases (STDs) as syphilis or gonorrhea. At the same time that the United States was persecuting Nazi's for war crimes against humanity, the grossest violations of civil rights were being undertaken after Franklin Delano Roosevelt's Good Neighbor Policy in Latin America.

Both Unit 731 and the Guatemala syphilis study were cases hidden for more than half a century, thus aptly fitting within Axelrod's sociobiological analysis. As all who participated were sworn to secrecy, the Japanese government consistently denied its existence, and it was not until the 1990s that the case became much more public. Interviews were held with its surviving perpetrators, now guilty for participating in wartime atrocities, because there had been no survivors as all victims had been killed in the war camps. A final report was issued in 2007, and it is shocking to see that none of the participants were ever persecuted or subjected to war tribunals.[175]

The Guatemala syphilis study also only recently came to light half a century after the event. In 2010, the medical historian Susan Reverby was undertaking research on the Tuskegee Syphilis study, working in the Thomas Parran archive at the University of Pittsburg. Upon her shocking discovery of its documents, Reverby then informed a former CDC head, Donald Sencer, who in turn shared it with the *Centers for Disease Control* (CDC). This information was ultimately relayed to President Barrack Obama, who then called Guatemalan President to apologize for the incident, which immediately turned into a media fire storm.

Both incidents had been closely guarded secrets in the public sphere, a notion by Jurgen Habbermas as the area of collective knowledge and public discussion, akin to the Greek agora.

Unit 731 was formed by Ishii Shiro, an ambitious medical doctor from the upper middle strata of Japanese society.[176] Studying at Tokyo University, he became regular visitor to the President's house—a practice which was shunned as he was operating beyond his immediate social circle, including that of his professors. Shiro eventually marries the president's daughter, a liaison which greatly aided his career. He went on a two year tour of the world's most important research laboratories, in particular Germany which at the time was a leading center in

---

[175] The Soviets after the war had gathered a great amount of evidence pertaining to these wartime acts, which the US refused to accept.
[176] His name can also be pronounced as Shiro Ishii, depending on the lexicon used.

chemistry and biology. The United States had been quickly catching up, and was at a relative par with European nations, even if it was less theoretically inclined.

Shiro's notion of biological warfare emerged out of various circumstances. He noticed that the 1925 Geneva Convention had banned the use of chemical and biological weapons, hinting at their enormous power and possible military benefit for Japan. Prior to WWII, Fritz Haber had invented synthetic nitrogen, bypassing the prior dependency on nitrates from Chile which were used for both fertilizer and explosives. The discovery and exploitation was personally tragic for Haber.[177] Ishii immediately realized that nontraditional weaponry might give a military advantage to Japan, and began pushing this agenda in upper governmental circles.

His decision to a degree can be accounted for within the context of Japan's recent military conflicts. In her industrial ascendancy, Japan had fought a number of wars with much larger nations, as the 1894 conflict with China. In that war, more men died by diseases than by bullets, in a ratio of 8 to 2; a fact which suggested the enormous benefits which could be gained by the incorporation of medicine into military practice. Shiro thus pushed hard for hygienic military tools, and at first developed a portable water testing kit which helped to prevent diseases as cholera. Its success could be readily observed in the Russia conflict of 1904, and helped to solidify the role of medicine in the Japanese military.[178] Shiro also introduced a portable water filter whose tremendous success helped win favor with upper level military officials.

---

[177] Fritz Haber also invented mustard gas. When his wife found out, she was horrified and issued ultimatum. As Haber could not back down, she ended up committing suicide.

[178] The importance of medicine to war was an emergent notion, brought to the fore globally by the work of Florence Nightingale, a nurse from England. Participating in the front lines of the Crimean war, she saw herself in a hygienic war, and her implementation of hygienic practices helped to drastically curb fatalities from 40% to 2%, a notion that would also be adopted by Bailey K. Ashford in Puerto Rico.

Shiro, however, increasingly saw medicine as a tool that could be militarily used for offensive purposes as well. His support in the military resulted in the creation of the *Epidemic Prevention Research Laboratory*—a name used to obfuscate its actual purpose and intent. Shiro and his new research team now participated in the active exploration of biological weapons for war, testing diseases as cholera, plague, typhus, and many others.

At the beginning of WWII Japan conquered Manchuria, and placed a puppet president, Henry Pu Yi, former emperor of Manchu dynasty. The area was renamed by the Japanese as Manchukuo, publicly claiming that it was an autonomous state— a pretense few believed. The purpose of the occupation was fairly obvious. Manchuria was an important source of raw materials for a growing industrial-military power, abundant in oil, coal and metals. While Japan might have had abundant human resources in expertise and education, she still lacked natural resources critical for an industrial state.[179] In less than three months, Japan had taken over the entire country.

Of strategic importance was the South Manchurian Railway, used to transport goods to and from Japan. Most of Ishi Shiro's research facilities were placed on the key hub of the region, near the city of Harbin. The first facility stood in the distant city of Beiyhinhe, a peaceful and isolated rural village, whose many villagers were forced to relocate.

At Beiyhinhe, Shiro and has assistants began drawing blood from test subjects not fully aware of the procedures they were being subjected to. In their experiments, blood would be slowly drawn from the 'patient' until the victim died. Ten cubic centimeters was the ideal amount. Generally speaking, the 'patients' were so weakened that they either died after the blood withdrawn or were simply too weak to fight back. At various points vivisections were performed.

The villagers grew suspicious about the facility; prisoners could be heard shouting in pain in the distance. A famous escape

---

[179] If Puerto Rico were a true industrial power, it likely would have also taken over nearby Caribbean islands long ago; lacking the human capital required for such an enterprise, the likelihood of military expansion in the region is relatively low—notwithstanding the existed of the United States.

by forty prisoners, planned for weeks by Li, led to public awareness of its local misdeeds. The escape reads like a novel, and were blessed with fortuitous circumstances. During a Japanese holiday, when the prisoners were given treats, Li knocked a guard on the head, took his keys and opened as many cells as possible. However, some prisoners were so weakened, that they could not escape. There was a rainstorm that evening, which knocked down the power generators electrifying a fence. Li heroically served as a human bridge at the bottom of a wall for others to climb, and was killed in the escape. Seven men made it to a nearby house, and told stories about the atrocities being committed in the facilities.

After the incident, Shiro decided that a new facility would be built, and closed the gaps which made the prior escape possible. The second Facility in Ping Fang stood in an urban area, closer to the city of Harbin. Its construction was rather particular, and distinctly made to control the flow of information: workers brought to the site had to wear head baskets. The nearby buildings also could only be one story high, so as to prohibit direct observation of the facilities, surrounded by a moat. The interior was composed of two parallel prisons, making an escape next to impossible. Thick concrete walls lined the jail cells. The jail cell openings where the blood samples had been drawn were now placed close to the floor, forcing the 'patient' into a submissive position during the procedure.

Paradoxically, the facilities were so well constructed, that they were difficult to destroy at the end of the war—and which served as evidence of its horrors. When the Soviets invaded at the end of WWII, the Japanese wanted to remove all evidence pertaining to the facility's actual activities. They killed all the remaining prisoners, first through gas, and, when that failed, with bullets. When the Japanese forces then tried to blow up the buildings, they failed. Today Ping Fan serves as a wartime museum and holocaust-like memorial.

It was clear that horrific crimes had been committed in its facilities. A number of diseases were used to infect the prisoners; the long list included bubonic plague, typhoid fever, and salmonella. Typically its doctors wanted fast acting diseases, as these would be the most effective from a military point of view. Cholera, for example, was problematic because it took 20 days before manifesting its infection. In the experiment, streets dogs were infected, successfully spreading the disease onto their owners. The bubonic plague, however, was the most successful, in that victims began dying after only three days—becoming one of the diseases most widely used by the Japanese.[180]

The prisoners used for these experiments typically came from the local population. Nobody had been fooled by Japan's pretension that Manchuria was an independent state, and hence there was a strong underground independence movement. Members of this movement were sequestered by the '*kenpeitai*', an elite police force who were all too successful in torturing 'confessions' as 'spies' or 'enemies of state' out of their victims. It is estimated that some 3,000 prisoners were taken, none of which came out alive. All had been used as human guinea pigs. The Japanese government claimed that Ping Fang was a lumber facility, and the term '*murata*' was a name given to the prisoners, which in Japanese means 'log'—an example of the inherent dehumanization of its victims by the Japanese.

The typical procedure was to infect the patients with a disease, to then vivisect bodies at different stages of the disease's progression. Japanese physicians obtained much more detailed information than they would have been able to otherwise.[181] In that the body is 'the sum of forces that resist death', its immediate dissection provided a detailed image prior to the point when pathogens began destroying the body, thereby capturing a series of accurate pictures of the disease's morbid impact. It would be this valuable medical information which would be used by Shiro as a bargaining chip when negotiating with the United States upon the war's end.

Ping Fang's victims came from all nationalities in the surrounding area: Soviet war prisoners, Chinese, Koreans, and so

---

[180] In the natural epidemiology of a new disease, the death rate is typically 33%, unequally distributed as occurred in Colonial Latin America with smallpox.

[181] Upon death, pathogens started acting on body, immediately damaging the available forensic data.

forth. Koreans were taken as translators, as these were trilingual in the major languages of the region: Chinese, Japanese, and Korean. When one Korean translator tried to assist the escape of a patient, he was summarily tortured, leading to permanent lung damage.

Shiro's research network was composed of 10,000 personnel. Shockingly, most of its practitioners were scientists from the private sector. It is estimated that these killed nearly a million persons in total (800,000 individuals). They key facility at Ping Fang experimented on 3,000 individuals, and was called 'Unit 731' upon creation of network of facilities. Other facilities included Xijing (Unit 100), Guangzhou (Unit 8604), Beijing (Unit 1855), and Singapore (Oka 9420). There was a deep and intimate collaboration with the medical sector who participated either by directly or indirectly by sending students. Ishii Shiro himself traveled regularly to Tokyo University where he presented the facility's findings, and his articles were even published in Japanese scientific journals. Ping Fang was an open secret within the medical community, as revealed in its very own journal articles. While the term 'monkey' was used in these, there were many clues as to its true character. The actual species of monkeys used was never identified, contrary to the regular practice of scientific paper. More strikingly, the recorded temperatures reveal that the subjects were human rather than simian, whose body core would never approach 104°F.

Each facility was dedicated to particular studies. Anda tested the explosions of pathogens. Prisoners were placed in concentric rings about a center holding a small bomb with the pathogen to be studied. Metal sheets were placed over the prisoner's torsos and light helmets over their heads. After its detonation, the radius of effectiveness of the pathogenic bomb was calculated. The explosion inevitably injured prisoner's extremities and limbs. The facilities at Xinjing were used for warfare on agricultural land, as the testing of pathogens for horses.[182] Pathogens as anthrax and

---

[182] Horses in the late nineteenth century was one of the main sources of transportation, akin to the car today.

ganders, which ate away at the body, were also tested, producing open wounds that never healed.

One of most horrific were tests were those undertaken by Yoshimura Hisato, a physiologist whom at first wondered why he had been asked to participate in a bacteriological facility. The Japanese military realized that the Soviets represented the most direct threat; prior conflicts held in cold Russian conditions had killed many Japanese soldiers. Hisato's experiments were thus designed to treat and cure injuries derived from extreme cold temperatures. Prisoner's arms were tied to a pole, and exposed to temperatures hovering 30°C below zero. Water would be sprayed on their arms to accelerate the tortuous freezing process; these were then tapped with a club. If they sounded hollow, the arm had frozen; tapping on the rigid extremity would often break it.

What is perhaps the most shocking aspect of all was how frivolous the series of experiments could be. No real medical knowledge was gained from them, the principal conclusion that raising the body temperature to 98.6°F was the best treatment. Hisato was never accused at a military war tribunal, and became President of Tokyo University after the war. Although he publicly denied being related to these experiments, in the classroom he would openly refer to his wartime experiences. In Japanese academia, going against superior is tantamount to professional suicide.

Yet it might be asked why were there were no war time prosecutions at all with regard to Shiro's medical network. One of most shocking aspects of the wartime atrocities is suggested by Hisato's case: those who participated rose to positions of leadership in postwar Japan.[183] Ishi Shiro died in 1959 from

---

[183] Kobayashi Rokuzu (President Of National Epidemic Prevention Institute),
Nakaguro Hidetoshi (President Defense Forces Medical School),
Naito Ryoichi (President, Green Cross),
Kitano Masaji (Chief Executive, Green Cross);
Kasuga Shinichi (President, Kenwood Electronics),
Yoshimura Hisato (President, Kyoto Municipal Medical University),
Yamanaka Motoki (President, Osaka Municipal Medical University),
Okamoto Kozo (Dean, Kyoto University Medical Sciences),
Tanaka Hideo (Dean, Osaka Municipal University Medical School),
Hosoya Shogo (Head, Infectious Diseases Research Institute).

natural causes, at the age of 67. It has been claimed that he even assisted the US army during the Korean War.

The circumstances in Japan towards the end of the war help account for this glaring ethical breach of justice.

While Japan had reached its maximum area of control, six months after Pearl Harbor, it would gradually be pushed inward. Although Saipan was identified as an important US base by the Japanese, its intended biological attack was thwarted before it could achieve its goal—and in the process helped to hide the existence of Japan's biological weaponry. The US was not aware of biological threat posed by Japan, but war conditions continually increased the likelihood of such attack. The losses towards the end of the war acted as a spurn to the use of biological weapons, if out of sheer desperation.

In Manchuria, the Japanese had dropped off wheat and corn laden with bubonic plague fleas. It was all too easy to attack peaceful village as Nigbo, whose contaminant bombs looked like a harmless smoke drifting in the air. The Japanese even developed a long submarine that could carry 3 war planes, which they had planned to use in an attack on California cities as San Diego. The US government was well aware of the determination of Japanese leadership. Collective appeals on behalf of the emperor for suicide were commonly abided by Japanese soldiers and civilians. The atomic bomb was used as an instrument of psychology, designed to place an end to such threats. Prior uses of napalm had the same destructive effect, but had not yielded an impact of the expected psychological magnitude.

Yet General Umezu Yoshijiro let it be known that Japan would not bomb US cities with biological agents, fearing that it would result in a merciless counteraction of global biological warfare. Biological attacks would also ruin Japan's reputation in the world, noted Umezu. At the wartime tribunals, Umezu never mentioned Japan's biological warfare capability. Dying in 1949, shortly after being sentenced to life in prison during the armistice, Umezu's was one of the war's tragedies.

After the war, various scientists were sent to Japan to asses its conditions and military capability. The team included microbiologists Murray Sanders and Arvo Thompson, as well as the chemist Norbert H. Fell. What is particularly striking is that Unit 731 sent men to become their translators, as Naito Ryoichi, who had actually served in Unit 731.[184] The relationship Naito developed with Sanders was one of scientific collegiality, of equal peers rather than of victor and conquered. Surprisingly, Naito was able to control the information that flowed to US military forces. Sanders was duped, and claimed that he did not suspect any biological warfare work had been done—but also that he was weary of his sources of information. Even though Thompson had been much more assertive and was able to determine that such efforts had in fact existed, he was unable to discover their full extent.

Incidentally, Sander's easy going manner allowed him to finally obtain the necessary information. It was well known that the Japanese were freighted of Soviet military forces, whom in contrast to the US occupied legions, had killed all Japanese soldiers they happened to come across. It was the Soviets who first to arrived at Ping Fang, forcing the sudden departure of its personnel. When fleeing on the first train leg, Shiro reminded his co-participants not to reveal Ping Fang's secrets to anyone, before taking a plane back to Japan. In his discussion with Naito, Sanders strategically warned him that the Soviets would not be as merciful with Japanese scientists as the Americans had been. It was at this point that finally Ishi Shiro, through the voice of Naito, used scientific information as a lever in his negotiations with the US government.

These negotiations actually ran at very high levels, and even top military leadership as General Douglass MacArthur, commander of the US military forces of the Far East, were aware of them. When MacArthur inquired into the quality of the information provided by Shiro, he was notified of its high value. In these deliberations, MacArthur was well aware that he did not need to give immunity to Shiro and the top leadership at all, in that he could obtain the same intelligence from lower ranking personnel that would not be subject to court marshal. However, it

---

[184] Ryoichi had also studied in the United States, and at one point attempted to get yellow fever samples 'for research purposes' from the Rockefeller Institute . His requests were wisely denied.

appears as if the Soviet presence itself also played a significant role in his decision making.

The Soviets had captured Japanese leaders and were making a big push to have these face war tribunals. North American authorities, however, replied that they did not have to hand over prisoners or share information, as its sharing of such assets had only been a friendly gesture. Ultimately, the Japanese wartime scientist were not prosecuted by the US government, as such trials would have produced data that inevitably fell into the hands of an even more powerful enemy, and possibly used against the US in future conflicts. Prudence was the upper hand of valor. Court marshal and war prosecution tribunals had their geostrategic limitations which had to be carefully weighed.

While the Guatemala Syphilis study (1946-1948) lacked the extreme violence which characterized Unit 731, it still shared many common elements. It principally affected captive populations whom were relatively confined and thus forced to comply with orders. The study occurred in a wide range of institutions, from prisons and psychiatric institutions to orphanages and army barracks. Another mutual feature was the sheer loss of human agency by all of its victims. Consent was never requested nor were the victims ever informed what the procedures they were being submitted to actually entailed. The treated population were identified as belonging to inferior groups, specifically descendants of the once glorious ancient Mayas whom now lived along a railroad line to the capital city in a poor area called "*la Linea*", characterized by a great deal of prostitution.

Finally, as in the case of Unit 731, information about its horrors did not publicly emerge for various decades after the incident. All of those involved actively sought to keep its activities and methods a secret, in part by reducing the number of people involved and by deceiving all other secondary participants—including nuns who assisted in the project. It is not clear what the level of complicity by the Guatemalan government has been. As in the prior case, we do need to analyze the role of medicine in its proper historical-military context.

The Second World War had raised awareness of the military cost of disease, particularly the amount of time lost to sexually transmitted diseases. As the character Bronn in the *Game of Thrones* series stated, 'soldiers make money to sleep with women.'[185] In total, some 350,000 new infections of gonorrhea occurred every year during the war, which was calculated to represent a loss of 7M man hours per year.

Sexually transmitted disease treatments at mid twentieth century were not particularly effective. These included silver proteinate and the use of calomel ointment to prevent syphilis. Their use was somewhat humiliating, as it became publicly obvious clear that the treated solider had an STD diagnostic. Salversan, a derivative of arsenic which was developed by Paul Ehrlich in 1907, was also used.

The problems caused by STDs became so serious, that a study was pushed forth by a leading scientific group at the *Office of Scientific Research and Development* (OSRD).[186] The study was performed by John Mahoney at the Terre Haute Center in Staten Island (New York) in 1944. It was a simple replication analysis, which attempted to infect prisoners with gonorrhea. Shockingly, it used taxpayer money to hire prostitutes previously identified with active cases of gonorrhea. The aspects of the study had actually been discussed for weeks, with regard to its legality and political impact. In spite of its morally problematic nature, it was recognized that the army had the legal capacity to undertake the study. Should information of the study become public, it was calculated that the political repercussions would be relatively low given the wartime context. A prior study done in England with similar methods had not been followed by a public outcry, so its political impact appeared to be negligible.

Mahoney, however, was very dissatisfied with the study's results. The rates of transmission were erratic, in that the experimenters could not achieve consistent rates of transmission required for definitive conclusions. After ten months, Mahoney

---

[185] The high amount of stress created in the life-threatening situations increases an individual's libido.

[186] The OSRD was headed by Vannevar Bush, a leading scientist with direct access to the President of the United States.

shut down the program. The proceding discussions recognized that wartime created special conditions which justified such studies, and would likely be regarded as inappropriate during peacetime. The nearing of the end of the war also forced their hands, weary of ending experiments which would likely not be allowed in the near future.

John Cutler, a recent graduate of medical school, was one of the participating physicians. He had a strange attraction to the study, which was characterized by the analysis of countless male genitalia. In light of the impending closure of the project, Cutler actively sought to repeat it in some location with a more lax public authority. Dr. Juan Funes, of Guatemala, pointed the way.

Funes had participated with Cutler in the Staten Island study, and suggested that the study be continued in Guatemala. Prostitution had recently been legalized, and all prostitutes were forced to undergo routine medical evaluations. Better yet, in 1933 a law (Article 10) had been passed which forced victims of syphilis to undergo medical treatment. Improving matters, soldiers in the Guatemalan army were known to often seek expulsion from the military by contaminating themselves with an STD.[187]

Political and social conditions in the nation also appeared to be positive for the study. Guatemala had been recently governed by the dictator General Jorge Ubico (1931-1944), who had taken steps to modernize the nation by bringing in electricity and telecommunications. During the consequent October Revolution, two sergeants came to occupy the presidency, the first of whom was Juan Jose Arrevalo (1944-51), a philosopher-reformer who created the Ministry of Public Health in 1945. The reforms initiated by Arrevalo were continued by his successor Jacobo Arbenz (1951-55), who enacted land reform given the vast

---

[187] They would take match and draw puss from another infected solider, applying the tip of the matchstick on their own genitals. In and of itself, the practice was puzzling, as no cure to the disease had yet been developed—as if one would opt to get cancer to cure a tooth ache.

unequal distribution of lands into enormous colonial Spanish haciendas.[188]

In the United States following WWII, many of the studies of the OSRD were transferred to civilian organizations, in particular the US Public Heath Service (PHS).[189] The physicians who had participated in programs as those as Mahoney's also gained positions within the new administration. The prospect of continuing the work in Guatemala was evaluated, along with thirty other projects. Cutler's project obtained both approval and funding, as the PHS had established a special 'secret' fund for projects as these. This secret fund would help it evade public scrutiny in that would not be included as part of routine administrative screenings, and thus narrowly restricted within the PHS's inner circle. The study was approved in April 1946, and its formal purpose was to verify the validity of various prophylactics in Guatemala, with a special emphasis on orvusmapharsen.

Its first funds of $110,000 went to the construction of facilities and laboratories in Guatemala. John Cutler, then 31 years of age, traveled in August of that year to begin operations. Upon arrival, he was presented to leading governmental figures, including the ministers of Health, War and Interior.[190] He would

---

[188] So backward was the nation, that it could be characterized by a certain medievality. Up until 1936, there still existed the '*mandamientos*', in Puerto Rico known as '*repartimientos*', in which Native American peoples were assigned to particular haciendas. The economic system was still characterized by debt peonage, whereby unfair exchanges turned native Americans into eternal slaves. At each unfair transaction, the Indian would lose money, eventually transferring the debt across various generations. In 1936, a law was passed, which was very similar to that of nineteenth century Puerto Rico was known as '*la libreta*'. The 1936 Guatemalan law required all men to be employed, and were also forced to carry a '*carnet*' (*libreta*) recording laborer information as hours and place of work. Guatemala sought to modernize and forge ahead early in the twentieth century, while still carrying the heavy legacy of its past, which limited its range of actions and policies.

[189] The PHS was formed in 1798 to look after the health of military men.

190 Dr. Carlos E. Tejeda (Chief of Guatemalan Army Medical Dept), Dr. Carlos Salvador (Director of the Psychiatric Hospital, or the *Asilo de Alienados*), Dr. Hector Aragon (Director of the *Hospicio Nacional de Guatemala*, an orphanage), Dr. Constantino Alvarez (Chief of Guatemalan

later claim that the contract he signed had given him a free hand with the patients of *La Penitenciaria Central* (prison), the *Asilo de Alienados* (psychiatric hospital), the *Hospicio Nacional* (orphanage), and the army barracks. Cutler assumed quite a great deal of power in spite of being such a young man, not unlike Ishii Shiro.

His study was morally and scientifically problematic from the very beginning. A total of 5,540 persons were studied. Some 1,308 were infected (or 'inoculated') with STDs, of which only 681 were treated (52%). It goes without saying that these procedures violated US protocol established during the Nuremberg Trials in Germany. The patients were never informed regarding their medical procedures, and formal consent was never requested. Psychiatric patients were not aware of what was being done to them, leading to explicit concerns by Dr. Arnold at the PHS. The same egregious treatment was also applied to infected children at orphanages, whom were often favored over adults. Children in the orphanages, as a rule, had no prior sexual contact, and underwent routine medical evaluations.

The cases are horrific wherever they were conducted. By far the largest 'inoculations' were undertaken at the army barracks and the psychiatric hospital, the most captive audiences whom had no choice in the matter, and could not leave even if they wanted to. Two particular cases aptly capture the nature of the experience. A soldier was inoculated in his genital area with gonorrhea. In spite of feeling a burning sensation, he was not allowed to leave, and was never treated. The case of "Bertha" at the psychiatric hospital is particularly tragic. About to die after having been infected with syphilis, she was subjected to even worse abuses given the terminal nature of her condition. She was infected with multiple STDs on their eyes and anus, developing puss before dying shortly thereafter.

Perhaps the worst aspect of the experiments is that they were scientifically sloppy. A baseline was never established, nor would the experimenters await the results of a given procedure before

Ministry of Health), Dr. Joseph Spoto (Chief of the Venereal Disease Division)

moving onto another experiment. It goes without saying that there was a troubling level of voyeuristic behavior by the physicians, who would observe the intercourse taking place with prostitutes, as if more focused on sexual act than the science underlining the study itself. Although 83 deaths occurred, their causes were never established. Finally, the highly funded project never undertook its principal aim: to test the viability orvusmapharsen as a prophylactic.

The entire project is even more puzzling when one considers that a successful treatment had already been discovered. Mahoney himself had been testing the use of penicillin for the treatment of syphilis. Initially tested on 4 persons , he went on to test its efficacy on 1,400 individuals, showing it to be a good antisyphillitic. There was even a presentation done before a packed audience at the PHS; the audience reacted with a 'thunderous applause' in 1944, and Mahoney's work was defined as one of the greatest contributions of military medicine.[191]

In other words, that an effective cure for syphilis had already been established prior to Cutler's study in Guatemala.

Mahoney had actually been well aware of the problems with Cutler's study. Just as Mahoney had previously, Cutler was having problems replicating the disease; the use of prostitutes simply did not work given the erratic infection rates. Mahoney himself had already hinted that Cutler should pack his bags soon. " . . . I wish you would give some thought to the future of the work in Guatemala."[192]

---

[191] Mahoney had not invented penicillin, but rather showed its effectiveness for this type of disease.

[192] These dynamics might also account for the Guatemalan government's collaboration in the study. Though speculative, it is likely that it had learned through Funes of penicillin, and likely saw the project as an opportunity for a type of technological transfer. Given that Mahoney had demonstrated the validity as a prophylaxis, its successful introduction into Guatemalan by the new administration would have been a huge political hit, helping to validate its progressive nature. Penicillin was then seen as a 'magic bullet', which was both effective and cheap to use. The laws of Guatemala do suggest a high probability that the incidence of STDs in that nation had drastically increased. Similar figures can be equally observed in Puerto Rico during the period, with syphilis rising from 1,900 to 15,203 cases. The disease in this case is serving

The role of information in the two preceding cases of medical abuse, Unit 731 and the Guatemala Syphilis Study, is clear. There was a strong need to control information about the nature of the unethical experiments by all of the professional physicians involved.

Information was controlled and monitored in many different ways in Unit 731. It can be detected in its architecture, in that no nearby buildings could be higher than one story. There was also information control via the translator (Haito) to Sanders. It is also striking that the Japanese government did not admit to the event until the decade of the 1990s. Once sanctioned, many of its participants provided ample testimony, some did so in an attempt to redeem old wrongs. It is important to point out that its victims could provide no testimony, as all had been killed in the installations during the war.

In the case of the Guatemala syphilis study, information was limited by placing control of study into the hands of only four individuals. In spite of its alleged scientific character, no formal report was ever published as a result. Cutler prepared a 1955 memorandum 'final study', which also never saw the light of day. The group never informed their patients of what was being done to them, but it was clear that something wrong had happened. In some instances, prisoners stopped collaborating, in that many blood samples were being drawn. Prostitutes also objected, in that they did not want to be part of a dehumanized medicine.

Such was the level of secrecy in both instances, that they might have easily escaped history and public awareness. Many historians in Japan actually claimed that the story of Unit 731 was a lie. Only through a lucky accident was the Guatemala case discovered by a North American scholar. Reverby prior to this had been arguing for the sanctity of the medical profession, but she now believes that such transgressions routinely exist in less developed countries.

---

as an indicator of broader social changes in society. Yet the level of complicity by the Guatemalan government is hard to determine. The available evidence does suggests a degree of ignorance. Dr. Arnold had commented that 'none would be the wiser' for the study.

We should not stray too far a field, however. The rates of Hispanic birth mortalities in Texas are so unusually high that they are the highest of any developed country. Rates of 35.8 per 100,000 (2010-2014) are alarming when compared to those in Italy (2.1), Japan (3.3), France (5.5), Chile (15.2) and even the United States (11.1).[193]

---

[193] The rate for the US varies greatly. Connecticut's rate is a low 3.53.

Minorities and the tragedy of invisibility

When Henrietta Lacks's body was autopsied, the paint on her toe nails was cracked and worn. Upon seeing this, her cousin Sallie immediately realized how much pain Henrietta had suffered, as she had been intolerant of unpainted nails in public. Worse still, her toe nail paint showed how quickly death had struck. Her health problems stemmed from the fact that her husband Dale had cheated on her, thus infecting her with syphilis and gonorrhea. The two diseases severely lowered her immune system, making her susceptible to a far worse one: the cervical cancer which killed her. The cancer had been so aggressive that white blobs of tumors were found displaced all throughout the interior of her body. The cancer had first been detected on January of 1952; by October of that year, Mrs. Lacks was dead.[194] Her last few weeks had been horrifically painful ones.

At the beginning of her illness, she had been sent to the Johns Hopkins University Hospital. During the first days of her stay, visitors had been welcome in her room, but were prohibited as her disease progressed. Her family would then stand outside her window to wish her well. At first Henrietta would look through the window, leading to such tender moments as when her hand would come to rest upon the window as a silent sign of love and tender affection. However, the pain became so unbearable, that she stopped looking and making any visible gestures on the window. Her physicians provided blood transfusions and morphine to reduce the pain and keep her alive, but eventually

---

[194] It is now known that cancer cells use the lymphatic system as a highway of transport, and is one of first things to be removed upon its detection. Unfortunately, the procedure leads to lymphodema, which leaks to the extremities, making one's arms and legs bloated and distended.

stopped as these had created an enormous blood deficit with the hospital's blood bank. Morphine no longer relieved her pain, and a pillow had to be placed into mouth so that she would not bite her tongue in half. As the pain continued to increase, Henrietta had to be strapped onto her bed, to prevent her from agitatedly shaking and moving. She would often lose consciousness in her state of inordinate pain.

These symptoms are perhaps understandable given the aggressive nature of her cancer, which was tragically unique onto itself. Cells as a rule do not live outside of the body.[195] When cells are cultured, they tend to grow along a two-dimensional axis in a Petri dish. Henrietta Lacks's cancer cells, however, grew in three dimensions. They were so resilient and multiplied so quickly, that they ended up creating a multibillion dollar industry, from which she and her family never benefited.

The original cell samples had been taken from her cervix during the initial evaluation. At the time there was a debate whether these cells were merely superficial or had crossed the cervical wall. Richard Telinde took the latter view and correctly argued that they needed to take more aggressive measures to save the patient. However, Telinde was moacked by his colleagues, which ultimately contributed to Lacks's rapid demise. George Grey ended up with the samples; realizing how unique they were, he went on television to talk about them. When pushed as to their origin, he claimed that came from patient 'Harriet Lane', a lie which made its source hard to trace. Grey, who also ended up dying from an aggressive form of pancreatic cancer, came to regret his decision. Much as Lacks, Grey shared her cells with the world, without obtaining much credit or recognition for his efforts, now referred to as "HeLa" cells.[196]

Their medical value resided in the fact that their external resilience allowed a whole host of studies to be conducted. HeLa cells reacted as any other normal cells to disease, physical or chemical stress. The cells were critical to the development of the

---

[195] One Nazi sympathizer played a hoax when they had allegedly produced a cancer culture from chicken heart cell.

[196] The name comes from first two letters of the first and last name, "HEnrietta LAcks".

Polio vaccine by Jonas Salk, saving thousands of lives.[197] Still today the sale of PENTActHIB amounts to $672M (2010). The sale of vaccines is a highly lucrative market in general.[198]

HeLa cells were also used to study possibility of space travel, at time when US was competing with Soviets to get to the moon. They were exposed to various gravitational forces to see if the human body could withstand the launch, contributing to the US lunar landing victory during the Cold War. The company Microbial Associates began mass producing HeLa cells, and typically sends entire crates of HeLa cell to facilities world wide.[199] Some 20 tons of HeLa cells are produced annually, and are currently used in more than 11,000 drug patents. Again, the Lacks family has never been a beneficiary of this financial largess.

The notion that Henrietta Lacks voluntarily "consented" to have her cancer cells removed is a highly problematic one. When individuals are in too much pain during a life threatening situation, there are visible changes in their state of mind. When one's life is threatened, the aim of cure and treatment will surpasses all other concerns. Under such conditions of duress, a physician requests a signature, most patients will willingly provide one, as all other concerns have become secondary at that moment. Worse still, not only did the family never receive compensation, but were constantly asked to donate blood by anonymous doctors whom seldom provided specific reasons for their requests.

It goes without saying that her family was rather traumatized by the long and painful saga. Not only is there a lack of

---

[197] Poliomyelitis used to be a very common prior to WWII in the United States. In 1952 there was an epidemic of 58,000 cases, of whom 3,145 died. It is estimated that there exist some 102 million survivors world wide. Many of these were turned many into paraplegics, requiring an iron lung to breath and live. Millions were obviously made from Salk's polio vaccine.

[198] In 2013, $24B worth of vaccines were sold. The *World Health Organization* (WHO) estimates the its market will reach $100B by 2025.

[199] Microbial Associates also accepted requests by individual researchers, and made millions of dollar doing so—to a degree that it displaced Tuskegee University as center of production.

confidence in white folk dressed in white gowns, but the family suffers from what has been termed "iatrophobia", or fear of doctors—which is common in the African American community. This distrust was of such a degree, that when Rebecca Skloot began to write her book about Henrietta Lacks, she did not think she would be able to finish it. Lacks's daughter Deborah did not know whether to trust Skloot, but learns she could and in the process the two uncovered another gross breach of medical ethics.

The two discover that Henrietta had a daughter Elsie, who unbeknownst to Debora had been taken to an insane asylum. After various attempts, they obtain Elsie's institutional records.[200] The insane asylum's physicians had been X-raying Elsie's brain, removing all of her spinal fluid to get better cerebral images. Although spinal fluid 'grows back' in three months, the procedure leaves permanent damage to the brain. A photograph of Elsie was found, whose grimacing face visibly showed the enormous pain she had been suffering.[201]

Afro-Americans perhaps provide the classical definition of a minority.[202] In spite of the fact that African Americans today make up around 40 million of the US population, or around 12.7% in total, they have higher rates of poverty than the average citizen. With an average income is $34,000, 25% live below the poverty level.[203] Forty percent of the population has been imprisoned, a figure which is noticeably disproportionate to their

---

[200] In spite of having been sacked, a janitor had preserved various insane asylum records for posterity.

[201] Skloot, a biologist by training, ended up writing a Pulitzer Prize expose of one of the worst breaches of medical ethics in the US, which helped expose the common use of African Americans in human experimentation in the United States.

[202] Brought to the Americas during the colonial period in what is known as the 'Mid Atlantic Triangle'. The triangle was made up of three nodes with three broad category of products: manufactured goods from Europe, slaves from Sub Saharan Africa, and natural resources from the Americas, which served as the raw materials for an industrializing Europe. A much larger number of slaves were actually taken to South America for the production of sugar. Brazil received some 3.65 M slaves, 'Haiti' (St. Domingue) 1.6M slaves, and the US 500,000 slaves, or 5.2% of the total exported to the new world.

[203] The median US income is $55,000

demography. Their rate of unemployment is double that the US average, 8.8% versus 4.9%. They have higher rates of children out of wedlock, 70% of whom live in single parent families. African Americans also typically suffer from a particular range of diseases, as hypertension and sickle cell anemia.

What role has medicine played in these dynamics? Are blacks victims of 'an inherent inferiority', as colonial slave traders would have us believe, or are they victims of history, as Steven Jay Gould claims?

The character of US colonial slavery is very different from its historical counterpart, as in Greece. Greek slaves were not socially stratified or biologically justified. A person could buy their way out of condition, and some of the most famous Greeks had been slaves, as Aesop. Greek slavery was not hereditary; a slaves children's children did not automatically become slaves. By contrast, African colonial slavery was a ruthless institution, systematic, exploitative, forming the nucleus of a retrograde southern US economy. It resulted in the establishment of a cultural value system with a very different scheme of values and social norms. from those found in the US North, and became the principal point of contention of the US Civil War. Would machines or men serve as basis of an economy?[204]

Yet, what do we mean by race?

Typically the term is used to refer to a particular set of traits associated with skin coloration. Norwegians are a 'race' because of their blonde hair and white skin, in contrast to the Nigerian race having kinky hair and black skin. However, from a scientific point of view, the notion is very crude. As Jared Diamond explains, social groupings can be based on a large number of criteria. If race is defined on a biochemical basis, one might actually en up placing members from two diverse subgroups next to each other, as Norwegians and Nigerians, in the same classification.

---

[204] Modern machines are the perfect 'slaves' in that they do exactly as they told, and can repeat an action infinitely, and have a lower 'cost of existence'. We today have become so dependant on machines, that we simply could not live without them. Imagine your daily routine if you had no car or bicycle, for example.

The typical notion of race is crude because the trait used is skin color. Melatonin is obviously a reflection of exposure to the sun; the more melatonin, the more 'protection' an individual receives. This protection, however, could be positive or negative, depending on the context in which the trait manifests itself. In countries of northern latitudes where there is little sunlight, dark skinned people do not produce vitamin D, hence creating a selective pressure for the lightening of skin.[205] The original migrating group to Europe might have been 'black', but selective pressure created by the environment was no different from that faced by moths in industrial England. Soot covered surfaces produced a new selective pressure favoring darker (black) than white moths, as the former were less visible to predators, and hence more likely to pass their genes onto future generations. Yet, we obviously do not attribute any pejorative internal change to the moth as a result. Moths are still moths, and do what all other moths have always done, regardless of their color.

African American were indeed subject to unique circumstances. The Mid Atlantic passage itself created new selective pressures, which in turn resulted in distinctive changes on the population's genetics. Those with higher levels of sodium in the blood tended to retain more water, and were thus conferred greater survival rates, passing their genes onto to future generations. Sub Saharan Africa also created the selective pressure of disease, as in the case of malaria, which in turn led to the development of the sickle cell, a curvature in the shape of the red blood cells which reduces the receptiveness and impact of the disease.[206] However, the net result is the balancing impact between the two dynamics, one positive, other negative. A positive net benefit of a harmful trait that confers a greater survival rate will, by definition, be passed onto future generations.

To some degree, we might actually claim African Americas are 'genetically superiority' for these traits. However, this claim is again false. As noted before, perfection depends on the

---

[205] The light skin allows for greater exposure, and in turn to greater vitamin D production in the body.
[206] The trait becomes detrimental when two recessives are combined.

environment in which a biological entity resides. As different environments will require different traits, no single set of traits is universally perfect or ideal. It is a rule which applies to both the diverse environments: disease, geographic, climatological, and biological . No single ideal form is suited for all, but are (again) relative to their existing circumstances. The notion of a 'better' or 'worse' set of genetic traits is facetious.

More importantly, slavery is a social institution which by definition is not open to the public sphere, and is similar to the Spanish hacienda. By definition, these social arrangements imply the loss of information, which accounts for the institution's propinquity towards abuses. Slaves by definition are the property of an owner, and hence subject to his will. Not subject to public scrutiny, slaves are invisible in the public realm, even when seen. The author Ralph Ellison captures the experience in his book *Invisible Man* (1952), named as one of the best 100 novels of the twentieth century. "I am invisible simply because people refuse to see me. . . . When they approach me, they see only my surroundings, themselves or figments of their imagination, indeed everything and anything except me."[207]

The most obvious examples of invisibility of Afro-American individuality exist in their use for exhibitions. The systematic comparisons akin to those in natural history were dehumanizing; the sanctity of human life was degraded when viewed 'objectively'. Two significant cases are particularly shocking.

Ota Benga Mbuti was a widower from Congo. All villagers, including his family, were killed during a raid, and Ota himself was taken captive and turned into a slave. Benga was 4'11" and weighed 103 lbs. He was sold to a zoo exhibit, and later obtained work in a tobacco factory. Benga eventually commits suicide with a handgun. The more well known case is that of the Hottentot Venus, Saarjie Baartman. Her large buttocks and sexual

---

[207] It is an odd historical fact that care of this property increased when insurance policies were retained. Seeking to minimize their outward payments to the insurer, the insurance company made sure of regular health inspections, so as to verify that the property was not being damaged unnecessarily. Ironically capitalism helped prevent extremes of abuse of the institution on which it was based.

organs were widely displayed in fairs, and made the subject of study and experimentation—akin perhaps to systematic rape. She dies at the age of 27 from alcoholism, a form of suicide. Both were not seen as individuals, but had been removed from their intimate social contexts.[208] Fortunately, Puerto Ricans fared better at the turn of the twentieth century.

It goes without saying that slaves were turned into food for fodder of medical practice. As their worth was based on their productivity, they became less valuable as they aged, and were consequently given to doctors for experimentation. Under these arrangements, if the slave died as a result of a medical procedure, then the doctor assumed their funeral expenses. If the slave survived, a couple of more years could be squeezed from him. There were a large number of cases of medical experimentation during the Antebellum period.

Dr. James Marion Sims in Alabama sought a cure for vesicovaginal fistula, or the tearing of the wall between anus and vagina during childbirth. The painful fistula was induced on slaves multiple times, without anesthesia. Sims eventually began providing morphine after the operation to induce sleep, but ended up turning his African experimental subjects into opiate addicts. The Phillip Morris Co. commissioned a painting which falsely portrayed him as 'saintly'. The painting has traits of a Norman Rockwell painting. The black slave sits peacefully on the table as Sims examines her. In fact, she would have been shouting, and would have to be held down during the procedure. Sims discovers that the use of silver forceps eliminates infection, and then travels to the North where he becomes a famous physician for his innovative practice. The beneficiary of African American experimentation were upper class socialites of the northeast, from whom Sims is able to raise funds to create a new woman's hospital.

---

[208] PT Barnum, creator of the Barnum & Bailey Circus, once noted that if skin could be turned white, racism would disappear in an instant. While this is doubtful, as Michael Jackson's whitening treatments attest, Putnam does point to the superficial character of the practice. The St. Louis Congress of 1904 paid individuals to actually go to African and get samples, presumably for display.

In a second set of experiments, Sims uses the slave "Sam", who was then relatively young at the prime of life. Sam suffered from syphilis, as 'all blacks are likely to have'. Sam's jaw was affected, but Sims' early gum operation fails, and turns Sam into an invalid. Sims later proceeds to remove his jaw, strapping Sam on a chair and held down by 10 men. Again, it is important to note that no anesthesia was used in the procedure. Mercilessly, Sims claims operation was a success, leaving Sam with a "permanent smile".

However, Sims is ultimately criticized by Nathan Bozeman, whom had been a former assistant in Alabama, noting that Sims never mentions African American experimentation in his articles. Sims statistics were also shoddy, as no true count of successes had been made. Sims uses his fame to discredit Bozeman. It will not be the only case of pandering popularity to attack one's scientific opponent.

Another beneficiary of medical slavery was Dr. Thomas Hamilton in Georgia. Hamilton studied the slave 'John Brown' for heat stroke, placing him in a steam bath until Brown collapsed. The chamber was not too different from torture devices; and allegedly uses it and other abhorrent experiments, to discover pills for sun stroke. In spite of their enormous success, they were a sham in that they were only flour pills, which were to be taken with cayenne pepper tea. The flour obviously offered no net effect, with the exception of the obvious rehydration which occurred upon the drinking of the tea. The spices in the tea gave the suggestion of 'active action', which were falsely attributed as a cure by Hamilton—a common psychological dynamic underlying humoral theory.[209]

Sims and Hamilton were but two cases of countless others, with varying degrees of severity and cruelty. James Dugas undertook experimental eye surgery on 80% black subjects, while Francois Proevots used slaves at similar rate (81%) for his studies of cesarean sections. Experimental ovariotomies and bladder

---

[209] Brown ultimately escapes to England where he publishes his famous biography, *Slave Life in Georgia*.

stone surgeries were similarly performed on countless African Americans. Their modern counterparts are perhaps no different.

The infamous Tuskegee Study between 1932 and 1972 was a crime of omission, in which physicians pretended to be treating 339 patients for syphilis when in fact they were only administering ineffective traditional treatments as vitamins and mercury. The African American's most important medical value was his corpse. A nurse was assigned to establish close personal bonds so that they would be directly taken to hospital immediately upon falling ill. The physicians wanted to see the direct impact of the disease on the body, performing autopsies no different from those undertaken at Unit 731.

The ethical violations of such medical practices are fairly obvious. Physicians were lying to their patients, who falsely believed they being treated, and hence constituted a core violation of the Hippocratic oath: 'do no harm'. In failing to effectively treat their patients, Tuskegee physicians promoted the development of the disease instead of hindering it. It should be remembered that penicillin had long been discovered, but was purposefully not administered to the Tuskegee patients. Surprisingly, the study had actually been approved by the US Public Health Services, the same agency behind the Guatemala syphilis study.[210]

The cover up is one more example of the human tendency to take 'any means necessary' to hide or prevent the public diffusion of information when gross unethical breaches have been taken by leading members of a community.

When suggestive hints of the study's existence began leaking into the public sphere, US Sentator Ted Kennedy immediately formed a commission to study the incident. Broadus Butler, a

---

[210] There is an interesting issue: if the rates of syphilis in other nearby counties were much higher than Tuskegee, why was it selected? (There was only 60% syphilis rate in Tuskegee.) It was chosen because it had good medical facilities. Booker T. Washington had formed the 'Harvard of the South' at Tuskegee University. With a grant by Sears Roebuck & Co Foundation, a good hospital was built, and initially designed to provide free medical service. Tuskegee was chosen simply because it was easier to hide autopsies, which were typically rejected by a community which sought to bury the body immediately upon death.

veteran and educator, thwarted the commission by mindlessly stalling its proceedings. He wasted time debating trivial issues, which limited the committee's effectiveness given the limited time available. At one point the commission even went to Tuskegee, and interviewed all the participants, doctors and patients, creating a comprehensive evidentiary repository of the event. In what has to be one of the most puzzling and perplexing actions by a congressional committee, the commission also ended up burning all of the evidence it had gathered, on a wholly spurious justification.[211]

In spite of the exogenous irrationality which overcame the afflicted committee, there can be no doubt with regard to the important role played by information in this case. All criminals fear the 'long shadow of the future,' which can ruin reputations for the rest of a person's life.[212] In the long run, it is tragic to observe how an elite institution, whose purpose was to serve a disadvantaged community, became wholly debased into an agency of repression.

Will the Tuskegee incident, or one similar to it, occur at the *Recinto de Ciencias Medicas* at the University of Puerto Rico; as it occurred already? It is difficult to answer. In a postmodern society, information is complex—at once ubiquitous and universally present, but at the same time often distorted. Cellphones and video cameras are everywhere, hence information on unethical behavior can be readily captured and diffused to the public at a moment's notice. However, hacking is pervasive, and digital information is also much more manipulable than traditional hard-bound storage forms. There is currently a great deal of pressure on the degradation of digital information systems for the threat they pose to criminal activity, whom use their artificial wealth to degrade the integrity of public information. Worse still, it is not entirely obvious when such invisible digital distortions occur. Most individuals have a false notion of security

---

[211] It was claimed that the destruction of evidence was done so as to protect the African American nurse involved, Eunice Rivers.

[212] It is highly unlikely that Bill Cosby, sentenced for raping women, will ever recover from the accusation. As in all of the prior cases studied, Cosby sought to prevent trial from ever being realized.

of the digital world, and the validity of the information contained therein. Younger people are perhaps much more sanguine, given the Edward Snowden revelations of 2013.

One of the most controversial issues pertaining the unethical scientific treatment of minorities is *The Bell Curve* (1994), written by Richard Herrnstein and Charles Murray.[213] The four year meta study of intelligence in United States created a firestorm of controversy when appeared, and was widely attacked by prominent scientist as Steven Jay Gould. If prior studies justified discrimination on the basis of the human body, now seemed to be based on an individual's intelligence quotient. Does the *Bell Curve* constitute a postmodern bias of racial prejudice and extend the legacy of colonial slavery?

The answer is complicated. It does represent the attempt to return to that golden age which was experienced in the immediate period after the Second World War, the US became the wealthiest nation in world history. While the book does contain a number of statistical errors, as a whole its general tenor pointed to troublesome tendencies in the North American republic. Some of its claims cannot be so casually dismissed, and does point to the decline of objective criteria in social evaluations. One of the negative features of the Civil Rights movement, ironically, are the dual standards it has created. African American requirements for college entry are a fraction of those demanded of Asian candidates, for example.[214]

While it might be presumed that the ethical violations of medicine and biology were more common prior to the emergence

---

[213] *The Bell Curve: Intelligence and Class Structure in American Life* (1994)

[214] The fall of standards is also visibly present in Puerto Rico, particularly in its private universities, which openly 'sell' diplomas to their students. The long term implication of such policies is the creation of a nonfunctional society. Architects will build houses that collapse, doctors will kill patients out of sheer ignorance, and plumbers will fix pipes that explode. The popularity and diffusion of the smart phone is also not necessarily a positive development. The process of reading requires deep concentration over an extended period of time. The hyperlinked mentality, on the other hand, looks more like the activity of an ADD child, jumping from topic to topic without much internal coherence.

of modern medical science, the historian of medicine Susan Lederer argues otherwise.

According to Lederer, the creation of bacteriology and scientific medicine created a noxious set of incentives which dehumanized patient-doctor relationships and led to a consequent increase in the violation of Hippocratic code of conduct. No longer was the patient someone to be cured, but instead became an opportunity for experimentation and a potential means of entry into the pantheon of science. The values of the scientific ethos which had traditionally characterized medicine were inverted, and involuntary nontheraptuic procedures exponentially increased. Rather than seeing abuses confined to specific isolated populations as minorities and psychiatric patients, the rest of the population became targeted as potential guinea pigs.

One common feature of both periods, however, remains the same: those with a low information profile are much more susceptible to being preyed upon by physicians.

Udo Wile at the Pontiac State Hospital in 1915 believed syphilis could be cured by inserting medication directly into the brain, and used a dental drilling bit to remove brain samples from terminally ill patients. Such were the horrors of Wile's gross ethical violations, that he was attacked by physicians on both ends of the ideological spectrum: William Keen (a conservative) and Walter Canon (a liberal). Given that tuberculosis was the AIDS of the era, a deadly disease without a cure, many physicians fell prey to the temptations of scientific fame. Doctors at the Vincent's Home for Orphans injected 160 children less than eight years of age with Robert Koch's tuberculin. Dr. Emmett Holt at New York Babies Hospital did so with a thousand babies, many of whom consequently died. Surprisingly, defenseless babies became common subjects for the testing of new vaccinations for diseases as scurvy, tuberculosis, and polio.

The rapid increase of medical facilities also provided common avenues of medical malpractice. In 1873 there had been only 178 hospitals with 50,000 beds, typically Samaritan places for the poor and the indigent. However, by 1909 more than four thousand hospitals with half a million beds had been constructed; helping to turn medicine into an industry whose growth far outpaced inflation during the twentieth century. Consequent legal battles in the public sphere over their regulation soon emerged during the twentieth century. Should animal experimentation be regulated? What about human experimentation?

While it might appear that the answer was obviously affirmative, the medical community was taken aback. Animals rights activists had quickly moved to establish well financed and coordinated organizations. By 1909 there were 334 societies, involving 42 M individuals, which led to 35,000 court prosecutions. Their growth and local successes in pushing municipal regulations led to more ambitious efforts at national regulation in Washington DC. These actions had taken the medical community completely by surprise and the response was overwhelmingly reactionary.

The movement inside the medical community was led by individuals as Keen, president of *American Medical Association* (AMA) and William Welch, President of Johns Hopkins, who tended to define any restriction as injurious to medical research. They pointed to the 1876 British Cruelty to Animals Act, noting that if it had been enacted earlier, Joseph Lister would never have developed the antiseptic. Animal research was defined as the *sine qua non* of medical research, without which it would not be able to advance. Such views had been influenced by the writings of Thomas Percival, who in 1803 argued that patients should never be told the truth about their condition, as it would be detrimental in their hope for life. A terminal diagnosis might lead the patient to become disaffected, and so the position was adopted as formal policy by the AMA in 1847.[215]

This stance was unfortunately later expanded to reject any regulation regarding human experimentation. It might be suggested that the issues became overly simplified when discussed in public forums, and the subtle nuances of each actor's positions were lost. Many physicians actually did believe that some sort of regulation on human experimentation was necessary, as there were far too many cases of abuse to suggest

---

[215] The policy was not revised in spite of studies that showed otherwise; families pull together in time of need, and their bonds of affection grow stronger under a crisis.

otherwise. A line had to be drawn in the sand, with regard to gross violations of medical ethics.

However, both Keen and Welch opposed any regulation whatsoever, incidentally involved in the Cornelius Roads incident in Puerto Rico. Both stances reflected a common philosophy: that of 'defending our own'. Unfortunately such stances established a horrific precedent, as if to publicly suggest that there were no repercussions to crimes in the field of medicine—or what is known as 'moral hazard' in the banking industry. Crossing moral boundaries would not result in any professional repercussions. One might ask the question as to why they won the legislative battle, when historical data suggested the contrary.

The medical community initiated an active propaganda campaign in favor of medical autonomy, whereby physicians were portrayed as martyrs and heroes who experimented on their own selves. The case most commonly used was that of Jessie Lazear. A doctor in Cuba during Spanish American War who experimented on the transmission of yellow fever, Lazear died with his body covered in black vomit.[216] Lazear had been in Cuba along with James Carroll and Walter Reed, the latter whom unexpectedly left for Washington DC during critical moments of the experiment. While Carroll also fell sick, he recuperated.[217] Lazear's story was repeated ad nasueum in a number of movies and novels, such as "Men in White" starring Clark Gable or "Yellow Jack" starring Robert Montgomery.

MGM also made a number of propagandistic movies about Pasteur, Banting and Best, in which the medical researcher is portrayed heroically, placing themselves in great danger so as to discover momentous cures which had eluded mankind for millennia. Their stories were oversimplified. However, that so many cures were actually found lent weight to such propaganda in the public imagination. Animal testing had proven beneficial to mankind, and the interpretation which became prominent at the

---

[216] When one researcher died, his wife threw herself onto her husbands body full of black vomit in attempted suicide. Yellow fever, however, is not contagious via this means.

[217] Incidentally, Reed dies in 1902, three years after the incident.

time was that any regulation on experimentation would be injurious to medical practice and to humanity as well.[218] The net result was that the landscape of legal oversight had become drastically weakened, and would come to establish the human rights context of the post WWII period. It had been falsely presumed by leaders that moral landscapes and moral culture was fixed but in fact, as any other cultural element, waxes and wanes over time. As each generational cohort has a distinctive set of experiences that are different from prior generations, they obtain their unique moral traits, which helps to account for the rise and fall of morality in America.[219]

We may conclude with a few discernable points arising from the case studies previously seen. The first is that individuals who fall outside the Habermasian public sphere tend to be subjected to atrocious medical experimentation. Such was the case for Afro-Americans in colonial period, whom as slaves did not benefit from constitutional rights. However this experience of medical abuse was by no means unique to them, and can be seen across many sectors of society previously noted, as psychiatric hospitals and orphanages.

The pattern of medical abuse accelerated with the foundation of bacteriology. Prior to it, biology was dominated naturalists, whom as a rule did not undertake much experimentation. This revolution, however, created a whole new set of noxious incentives. In short, the temptation of medical fame, created a conflict of interest between the attending physician and his patient. Was the patient someone to be cured or someone to be experimented upon? The temptation was too great for some physicians and the rate at which medical abuses occurred increased. New opportunities for malfeasance as the rapid

---

[218] It is perhaps equally important to point out, that the counterbalancing forces had also disappeared from the scene as the century progressed. The multiple rights organization early in the century had disappeared; many of its scientific leaders had died as the case of Walter B. Cannon who ironically fell to stomach cancer caused by the radioactive substances used in his scientific work.

[219] Moral stances need to be constantly 'protected' to insure that ethical standards do not decline over generations.

expansion of hospitals led to the dramatic rise in actual cases of gross negligence and malpractice.

Regrettably, the ego of medical men impacted the political process, subverting the role and function of the law. Rather than creating the most reasonable social order, where both animal and human experimentation would be regulated, the stance was wholly rejected by Keen and the AMA which he represented, creating a noxious precedent for the rest of the century. One of the most impacted populations would be minorities and individuals with uncertain positions in the public sphere of American life.

Henrietta Lacks was one such example of many others in the African American community.

Fraud in Biotechnological Research

In 1985 Margot O'Toole was a post doc in Thereza Imanishi-Kari's group at MIT, and found what she believed were serious discrepancies with the work being done in the laboratory. The work had been published an article in the journal *Cell* (1986), co-authored with David Weaver, Moama Reis, and David Baltimore, a Nobel prize winner. Weaver was then a student of Baltimore's, while Reis was a Brazilian postdoc working in the same laboratory. When O'Toole confronted Imanishi-Kari, the director denied the allegations.

After verifying the work and once again getting no results, O'Toole then turned to university administrators, in particular Henry Wortis, who had helped her get the position in the first place. Wortis had been her thesis adviser at Tufts, and was currently evaluating Imanishi-Kari for a tenured position. The allegations were serious and potentially implied grave repercussions to Imanishi-Kari's career. Tufts found no evidence of wrongdoing. O'Toole had also turned to MIT leaders as Herman Eisen, who also did not find any wrongdoing. While Eisen did write up a memo, it was not shared with anyone else.

When Imanishi-Kari found out about her assistant's activities, she fired O'Toole from her postdoctoral position at the laboratory. A professorship that was also going to be offered to O'Toole at Stanford by the Leonard and Leonore Herzenberg suddenly vanished. At the beginning of her scientific profession, O'Toole suddenly saw herself with the possibility of being without a career or future in science. Unbeknownst to her, however, the case had quickly escalated.

In her inquiry of Imanishi's work, O'Toole had gotten in touch with Charles Maplethorpe, another postdoc who had previously worked in Imanishi-Kari's lab—and whom also had serious reservations regarding the work being carried out there. Unbeknownst to O'Toole, Maplethorpe then spoke to two science investigators, Ned Feder and Walter Stewart, at the *National Institutes of Health* (NIH) about the issue. The two had been previously involved with cases of scientific fraud, specifically that of John Darsee which had obtained a degree of public attention, and which had made Maplethorpe aware of their activities. There had been no doubt regarding the allegations of fraud in Darsee's case.[220]

As Feder and Stewart began digging into O'Toole's allegations, the case quickly escalated and resulted in a congressional committee inquiry by John Dingell Jr., the powerful chairman of the House Energy Committee. Dingell Jr. was a Democrat whom was used inquiring into all sorts of wrong doing in government; in contrast to other legislative powerbrokers, he went after the major players. He sought to get to the heart of corruption in the modern United States, an attitude which may be accounted for by his family history.[221]

---

[220] Darsee had been fabricating data since his days at the University of Notre Dame, and 8 of 10 the papers coauthored at Emory were forced to be withdrawn by the NIH team.

[221] Dingell Sr. had been a Michigan representative, an Irish migrant whom had the fortune (or misfortune) of unexpectedly gaining a seat in Congress. During the Great Depression, he was forced to leave school to begin work to help

In the meantime, Nobel prizewinner David Baltimore had been elected as President of Rockefeller University as news of the allegation slowly began entering into the mainstream media. Various important scientists at Rockefeller concurrently resigned in protest from the University, taking with them huge research teams—sometimes as large as 30 individuals per team. While the particular scientists alleged that their motivation had not been influenced by the emerging Baltimore scandal, it was hard to believe such claims. Baltimore's own mentor James Darnell, also resigned from position as vice president of Academic Affairs, which ultimately precipitated Baltimore's own resignation. To say that Baltimore was 'humiliated' by the incident would be to make an understatement.

The O'Toole, Imanishi-Kari, Baltimore case lasted a decade and led to a series of reforms in the regulation of the US scientific community, specifically designed to prevent such abuses of power from occurring in the first place. Some of the changes were more were more substantive than others.[222]

After 1996, scientific journals implemented a much stricter criteria of authorship. All authors are now required to sign a document stating that they have read the entire manuscript. Many important senior professors had been singing off on research to gain credit without actually verifying the scientific work

---

support the family, initially distributing newspapers, forming a union at the newspaper. Caching tuberculosis, he travels West to improve his health, where he met wife. He later moves to Detroit with his two sons, both of whom study chemistry. Dingell Sr. heard of Franklin Delano Roosevelt's call for new leadership in congress, and when he ran for a spot he miraculously won. Dingell Sr. then pushed forth a number of leading medical reforms, which combined various institutions to create the NIH. He also tried to get universal health care established, but was unsuccessful. On a routine asthma check at the Walter Reed Clinic, Dingell Sr. mysteriously died from a heart attack.

As Alexander the great, his son fought off contenders and was able to emerge to occupy his father's position. Dingell Jr. at 6'3" was a big guy, and much to everyone's surprise, his leadership lasted a decade, from 1986 to 1996.

[222] The *American Society for Biochemistry and Molecular Biology* created an ethics code in 1997, which in fact was rather hollow as it had no procedural meat.

described in the paper. Journals also began requiring notices that the images used in scientific articles had not been tempered with. Any institution receiving NIH funds was now required to establish formal guidelines, specifying levels of responsibility; they were being forced to develop stricter procedures with regard to scientific abuse.

Universities were also required to develop a stricter retention policy for the collection, storage, retention of data, so as to prevent data from being tampered with after the fact. Such policies insured that the data used in a particular article would be accessible to other sciences for verification; in other words, scientists were now required to share their data with other researchers as well. Finally the events coincided with formation of the Office of Scientific Integrity, today the Office of Research Integrity.

One of the prevalent questions of the case pertain to its escalation. Why did it explode as quickly as it did? Even James Watson, who had been asked in the media about it, was perplexed by its surprising growth. Why did it explode as it did in 1986 and not earlier, say 1966? Was the case merely one of scientific misinterpretation or one of scientific fraud? More importantly, what does it tell us about the vulnerability of young scientists when they come across improper behavior by senior scientists with much more power, network of contacts, and prestige?

A little bit of history might help to clarify the matter. Ambiguity is an integral part of science, and much of a scientist's effort is undertaken simply to clarify the source of such ambiguity. An interesting case is that of Peruvian bark.

Peruvian Bark, also called Jesuit's bark or cinchona, was discovered in 1630. Its actual function and role was not well understood during the seventeenth century, as it crossed across many different medical theories: iatrochemical, humoral, fermentation, and so forth. At one point, its potency was accounted for its 'hot-dry' bitter taste, but by humoral standards it should not have worked with intermittent fevers which today we know as malaria. Several thousand people were treated in 1653 by Cardinal Juan de Lugo in Rome, who tested it and found it to be most beneficial. However, its benefit was not necessarily taken to be taken for granted, as we would today.[223]

At various points of the debate on the effectiveness of cinchona, scientist used their social status to squash rival theories. Jean Jacques Chiflet used it to treat the governor of the Spanish Netherlands; when his important patient relapsed, Chiflet wrote against it. Given his prominent position, the bark was then rejected and discredited in Paris. One problem is that the bark was often mistaken for other barks. George Ernst Stahl also opposed Peruvian bark in Germany. He observed that since vomit was an important process of medical recovery, given its expulsion of noxious substances from the body, as the bark prevent vomit, Stahl concluded that its impact on the healing process was detrimental. Again, the claims made by a famous intellectual led to its long delay as well.

Hands on experimentation, however, shifted opinion in its favor. Jacques Minot did "exterior experiments" with Peruvian bark on blood and milk, observing that it helped to preserve substances outside of the body. John Rushworth in the colonial United States found that the bark saved a patient with gangrene, and wrote letter to Hans Sloane in 1731 describing his finding. Sloane, then the King's surgeon, tested it and found it to not only prevented gangrene, but to generally improve the patient's condition. It helped to restore a patient, and in turn to a successful amputation. John Pringle, general physician to the Majesty's forces, rubbed it onto various substances, to see if it affected the putrefaction of meat, egg yolk and so forth, using sight and smell as indicators. For Pringle, as putrefaction was the dissolution of parts, he noticed that the Peruvian bark had a positive antiseptic effect.

It goes without saying that ambiguous situations where evidence supported both claims and counterclaims, the power and

---

[223] A lot of information emerging from the Hispanic world was false, as the case of the peccary. Its fatty structure atop its neck was believed to have been a nipple or a breathing hole. Dissection by members of the Royal Society discovered it to be a scent gland. Hernandez claimed that bird laid eggs underground that had no yolk, were enormous, and that only one such egg was laid—claims which went against common sense.

influence of an intellectual's social status helped to formally establish the acceptance and validity of an idea. An idea promoted by powerful and influential leaders will be much more likely to be adopted than by less influential colleagues and competitors.

To show how ambiguous the interpretation of the Peruvian bark could be, we may also compare the preceding views with those of Thomas Willis. Willis abided by the school of iatrochemical medicine. When beer was fermented, its cauldron became hot. The taste of blood, as either acidic, salty or bitter, also provided indication of its internal chemical processes, and hence of nature of the illness at hand. Willis believed that Peruvian bark worked because it initiated a new process of fermentation in the body, and referred to it as a specific. Thomas Sydenmham later believed that the Peruvian bark was the only specific in medicine, as an antidote specifically for particular 'venom'. The relative scientific fame of both medical men helped to push the Peruvian bark into general acceptance in the Western medical Canon.

During the nineteenth century, Louis Pasteur believed in the 'biological exhaustion theory' or the notion that a virus produces its own antivirus. It would somehow consume in the body that which eliminated its own risk at a later date. Although a mistaken theory, it led to particular therapeutic procedures. Pasteur would begin treatment with high doses and then move to lower these. But his patients provided conflicting evidence of the theory's validity. Henry Toussant, a constant critic of Pasteur, provided his own chemical theory which turned out to be correct. Pasteur also clashed with Emile Roux, who had originally been a physician in the army.[224] Roux as a physician was much more sensitive to injuring patients, and strongly opposed some of Pasteur's wanton clinical experimentation.

In one case, Pasteur treats a boy using the underlying biological exhaustion theory, but the boy dies soon thereafter. The father was going to sue the institution, but the political

---

[224] Roux had been kicked out of the army, and as O'Toole did not have many career prospects early in his profession. Through other student of Pasteur's, Emile Duclaux, Roux able to obtain position in Pasteur's installations.

powers that be prevented the lawsuit from ever getting to court. Pasteur by then was powerful enough to significantly reduce the cost of negative medical outcomes. It is also to be noted that Pasteur was more focused on science than therapeutics, hence his greater patient disregard than Roux. However, Pasteur eventually turned to chemical theory at a time when its exponent, Toussant, had already died. Pasteur adopted Toussant's ideas, and inverted his own prior treatment, using low rates of serum, which turned out to be much more effective than his earlier procedures.

It is important to note that Roux had not participated in the fatal experiment, and had actually warned Pasteur that such a treatment would injure the patient. These debates, however, do not appear in the record. Pasteur was in fact legally forbidden from injecting patients, as he was not a physician. Roux, however, chose not to 'betray' Pasteur, and does not denounce him to authorities or publicly expose him. Although Roux was clearly the junior partner, his attitude is somewhat puzzling. After Pasteur's death, Roux not only defended his mentor, but proceeded to lionize him, turning the figure into a type of cult worship.

Roux's response to the ethical dilemmas he face can be clearly defined as a conflict of interest. He was clearly seen as Pasteur's intellectual descendant, and stood much to gain by his implicit association with Pasteur. In short, Roux financially and professional benefited from Pasteur's posthumous fame and reputation, in spite of his mentor's wrongdoing. After Pasteur's death, Roux never mentions the events in his memoirs. When personal interests have a particular favorable outcome, they stop being an unbiased neutral observer.[225]

The case of Roux-Pasteur might be contrasted to that of Charles Banting and Harold Best, Banting's postdoc. Their collaboration resulted in the discovery of insulin, via a rather uncertain and tortuous route. At one point, the two almost came

---

[225] The same occurs when drug patent trials are undertaken by a drug maker. There is a blatant conflict of interest in that such pharmaceutical corporation stand to make billions upon the successful testing of a drug. Since they want drug to succeed, they will make trial 'succeed', and hence constitutes a conflict of interest which dominated the early biotechnology industry.

to physical blows.[226] Banting had been in a precarious financial position for some time. He was so low on funds, that at one point he sought the aid of his former professor Henderson.[227] Banting also confronted his own immediate boss John James MacLeod, demanding better working conditions, including the replacement of the irregular laboratory floor. MacLeod rejects all requests, and informs his subaltern that "as far as Banting was concerned, he (Macleod) represented the University of Toronto."

It is clear that Banting greatly needed Best's assistance, and so the social division between two was not large. Both worked in the same office space, and could blame nobody else but themselves for any experimental failure. Banting in fact had made a great sacrifice to obtain his discovery. He had left a comfortable position in London to return to Canada. Prior to the rise of biotechnology, all insulin was based on his discovery, and was extracted from animals.

We might suggest that drastic polarities of power between mentor and student are as ruinous to their relationship as all other social hierarchies. Individuals vested with a great deal of power tend to abuse such power, as in case of MacLeod's treatment of Banting, or of Pasteur's of Roux. These relations might be contrasted to the democratic traits which are often presumed to underlie scientific exchanges, which in turn has implications for ethics. Obviously mentor-student relationship is not one of equals, and thus are characterized by a particular 'bias' against the student's claims and activities.

We may point out that all behaviors are shaped and guided by particular social 'walls', as those produced by legal limitations, which constrain behavior. Such limitations are not necessarily a good thing, however as external imposition of rule without full understanding of the surrounding circumstances are a recipe for

---

[226] Banting had accused Best of being sloppy. Best, coming from an upper class family, clenched his fists and was ready to get into a fight. However, he turned around and stays up all night, redoing his entire work, getting much better results. Banting had been right all long, and the two end up becoming close friends in the end.

[227] Banting slammed 7 cents on the table, claiming that was all he had to live on—leading Henderson to promise a future job.

injustice and unethical outcomes. It goes without saying that all individuals are caught in a web of relations, with divergent series of ranking amongst its members. Any single individual might have a higher ranking in a particular context, but a lower one in others.[228]

These realities have important implications for the social dynamics of science, particularly biology in the United States during twentieth century.

As science has shifted from 'little science' to 'big science'—with its high funding, complex technologies, and large teams—social hierarchies have been introduced, which erode the notion of peership in the community. These social structures present challenges to the activities of science when improper conduct is detected. The notion of a Nobel Prize also alters scientific peership, in that its recipients are 'worth more than others', clearly demonstrated in the O'Toole-Baltimore case. The Noble Prize becomes a double edged sword in such occasions. While the scientific community might want to recognize substantial scientific contributions and does not intend to make tyrants of its recipients—such unfortunately have been its results.[229]

The organizational, financial, and technological changes in biology throughout the twentieth century have thus had an unfortunate byproduct: the substantial increase in fraud. Prior to World War II, cases of fraud were relatively rare in both biology and science, generally speaking. Such cases were typically nothing more than embarrassing exceptions to the norm. Student typically would work with a respected researcher, and when evidence of flaws or wrongdoing were discovered, it only degraded the status of student and did not necessarily have glaring career consequences—or legal ones for that matter. The student simply had not complied with the high ethical standards of the time.

---

[228] Those whom we might presume to be big fish in a particular context, can suddenly become very small fish when the social context is shifted.

[229] Pharmaceutical industry has unfortunately used legal constraints to justify thievery.

However, after World War II, the number of cases of fraud began to rapidly increase. Some cases had been of an involuntary nature, due in part to publication pressures. When a 'wunderkind' appeared with studies which validated his mentor's theories, the mentor was all too ready to accept them as 'important', and worthy of publication. The mentor's vested interest in his theory created conflict of interest in his relationship to his protégé.

Gradually, however, as the rates of scientific fraud increased, so did the incidences of purposeful deception, as the presentation of false credentials. In one case, a student pretended to hold a non-existent masters from the University of Cincinnati. The case of Vijay Soman at Phil Felig's office in Yale was particularly shocking. Felig had been asked to review paper, which he rejected but then showed it to his student. Soman in turn took the paper, made up evidence for it, and submitted it for publication. In an ironic twist of fate, the authors of the first paper received Soman's article for review, and immediately denounced the gross plagiarism. Felig, who at the time had applied for a position at Columbia University, was immediately dismissed. Yale was extremely slow in dealing with the issue. It took so long to resolve the issue, that by the time the enquiry had been completed, Soman had graduated and was already working back in India as a tenured professor.

The great deal of collegiality in the sciences, and in academia broadly speaking, has a detrimental aspect. When pervasive, professors take each other's words at face value, and thereby ironically fail to apply the same rigorous standards they apply to their own research. Had the participants in the prior case studies simply verified academic credentials or routinely evaluated their student's work, many of such gross violations would never have occurred in the first place. Collegiality is mistaken for shared standards of rigor and moral integrity between peers.

During the formation of the Office of Scientific Integrity, there was a clash over this very issue between Robert Charrow, then Deputy General counsel of the HHS and James Wyngaarden, NIH director. Charrow believed there was too much collegiality in that scientists routinely failed to show enough of a critical attitude in collegial exchanges. Charrow thus wanted more of a 'judiciary' character to the OSI, specifically a separation of powers between an independent panel which would evaluate the facts of a case and a separate adjudicatory committee to assess its punitive consequences. In the process, a situation of potential conflict of interest would be avoided.[230] In Charrow's model, the accused would also be given the typical legal rights afforded in a court of law: the ability to confront an accuser or the ability to see the evidence used against one from the very beginning of the evaluative process.

Wyngaarden's views were the complete opposite. He wanted to retain the collegial aspect of academic relations intact, and as such called for no separation between the two branches of a case—the investigative and the adjudicative. Ironically, however, in Wyngaarden's legal framework, the accused would not have the ability to review evidence or confront his accuser, a feature which was justified on the basis that it would avoid a direct confrontation between the two parties. Tragically, Wyngaarden's framework succeeded.

Charrow was very skeptical of the OSI to be established, but strangely was placed in charge of its institutional creation. His failure to act led to his 'pocket veto'. However, Charrow did give the institution is formal name, which to him had an Orwellian character. Charrow did not implement the institution until he resigned from his charge, and thus delayed its formation. Charrow was ultimately correct in his assessment. The number of cases of scientific fraud had increased so much, that Al Gore Jr. was going to form a commission for evaluation within the Energy and Commerce committee, so dominated by Dingell Jr. before his entry into presidential politics.[231]

In fact, during the early 1980s, transnational corporations were beginning to invest heavily in science, as the promises of

---

[230] If the judge of a court case is the same policeman who arrested an individual, the case will inevitably result in a validation of the accusation.

[231] What is striking is that Al Gore Jr. abandoned this important project, only to become one of the most inconsequential vice presidents in US history. While he might have helped Bill Clinton snatch the presidency, one has to wonder what would have happened had he formed a commission to investigate scientific fraud in the US. It is certainly the case that impact would have large.

biotechnology seemed golden. Millions were invested in universities and in biotech companies, and their entry coincides with the rise of scientific fraud in the life sciences. It is in such a setting where the Dingell committee hearings were set: an increase in the scientific fraud that was occurring more generally, denounced by a young female scientist who had few safeguards against more senior male officers. Such were the principal features of the Baltimore Case.

There had initially been a great of hope for the *Weaver et al* paper, which was defined as an extension of Baltimore's work. At the age of 37, Baltimore had won a Nobel Prize for groundbreaking work with Howard Temin at Renato Dulbecco's office, all whom shared in the award. They discovered reverse transcriptase, and in the process overturned the existing paradigm in biology, specifically the notion that information flowed only one way, from DNA to proteins. Reverse transcriptase showed that information could also flow 'upwards' to the DNA as well. Baltimore and Temin had studied viruses which induced cancers by modifying their cell DNA, and it was hoped that their discovery would lead to new miracle cures for seemingly incurable diseases as cancer, AIDS and so forth.

The potential conflict of interest can be immediately detected in the now infamous *Weaver et. al.* paper: a student (Weaver) was validating his mentor's theories (Baltimore).

The study allegedly proved that transgenic mice would adopt foreign DNA, which in turn would allow physicians to force the immune system to target particular cells. All autoimmune diseases as lupus are odd, in that the body fights its own body, thus leading to a series of well known medical complications. The particular issue challenged by O'Toole pertained to the Bet1 reagent. Two reagents were used in experiment, specifically AF6, which identified the control mice, and Bet1, used for the transgenic mice. Imanishi-Kari's *Cell* paper concluded that that Bet1 was unique, in that it only coded for a particular variant of the antibody, and was hence a consistent marker of transgenic mice.[232]

When Margot O'Toole first began working in the laboratory, she assisted Moema Reis in her work. The Brazilian Reis was favored by Imanishi-Kari for her compliant hard work; Imanishi-Kari herself was also Brazilian.[233] Reis, however, leaves the laboratory after accepting a position in Brazil. Imanishi-Kari had been visibly upset that Reis had left. This institutional vacuum created a new opportunity for O'Toole, who strangely was unable to replicate many of Reis's experiments, which O'Toole found perplexing at first. She was told by Imanishi-Kari that it must be O'Toole's own fault. However, when O'Toole then tried repeating the experiment multiple times, she was never able to find the expected results. Her main conclusion, which was the correct one, was that Bet1 was not uniform in its function, and coded for both original and transgenic mice. If true, it invalidated the entire *Weaver et. al* paper: the claim of its unique coding was plainly wrong.

O'Toole then requested to personally see Imanishi-Kari's original notebooks, so as to look at the original data on which the scientific paper had been based. Imanishi-Kari refused to do so, which O'Toole found particularly disturbing. After much searching, she came across Moema Reis original notes, some 17 pages of the raw data on which the *Weaver et. al.* paper's conclusions had allegedly been based. These pages would be critical to the formal congressional investigation later on. O'Toole immediately realized that the paper would not stand.

She did not initially claim that fraud had been committed, but rather it was clear to her that a retraction from the journal *Cell* was necessary as its conclusions were scientifically unsound.

---

[232] IgG is most common (75% of all human), while IgM was largest but the least prevalent (10% of human). This in turn had two variant, mu^b (control mice) and mu^a (transgenic mice). Imanishi's allegation was that Bet1 always coded for mu^a, which is to say that it was claimed to consistently identify transgenic mice.

[233] Imanishi-Kari came from a peasant Japanese family; she had had struck on her own, studying in Sao Paulo and Tokyo before obtaining her PhD form Helsinki University.

However, Imanishi-Kari's behavior was more troubling in that she began claiming in other articles that O'Toole had found evidence supporting the paper's original claims, which was an obvious misrepresentation. O'Toole was naturally upset by Imanishi-Kari's expositions, as they were portraying a false image to the world about the laboratory's interior dynamics. It was at that point when O'Toole went to MIT's office of academic affairs to complain to Eisen.

Herman Eisen was then head of the MIT immunology department, and had originally recruited Imanishi-Kari. O'Toole's claims placed him in a conflict of interest quandary; if true, it would reveal poor criteria on Eisen's part—and, in turn, affect his own standing at the institution. O'Toole also contacted MIT Ombudsman Mary Row, who was then an assistant to the MIT president as well. O'Toole also consulted Bridget Huber, a friend of both parties; Huber in turn spoke to Robert Woodland, a professor of molecular genetics at the University of Massachusetts Medical School—a true neutral voice as far as all the actors were involved. Woodland realized that the affair could explode into a major controversy, and recommended O'Toole consult the issue with Henry Wortis, her thesis mentor at Tufts, who was at time undergoing review of Imanishi-Kari's candidacy.

The general reaction by the various university committees was simply that nothing had really happened. In her response, Imanishi-Kari claimed that the conflict stemmed from a personality issue. To her, O'Toole's dedication to science was questionable given her expressed desire of motherhood; O'Toole had not applied to the necessary grants while at the laboratory. As a result of O'Toole's apparent lack of interest, Imanishi-Kari placed her cleaning mice, which implied O'Toole did not have necessary competence for her role as postdoc in the lab, a move which had also soured their working relationship.

O'Toole also challenged David Baltimore, who arrogantly noted that he had met many troublesome people like her before. She had the option of presenting her claim to a scientific journal, but in turn he would write a rebuttal. As a Nobel Prize winner, his word would inevitably weigh more than hers, so, implicitly stated that she should not bother trying.

O'Toole was naturally very depressed and weary with the emerging affair. She could not believe what was happening, but became more reassured when she spoke to Maplethorpe. Maplethorpe himself had told friends of political intrigues in the lab, but most refused to believe him given the reputation of the individuals involved. Similarly, when Feder and Stewart began inquiring into multiple incidents of scientific fraud, they had obtained a 'bad reputation' in the community. Early drafts produced by them of the Darsee case had already resulted in libel threats.[234]

O'Toole was at first furious that Mapplethorpe had informed them of her quandary, as she only wanted a retraction of the faulty paper. Her charges were not of fraud and she simply did not want to cause a major controversy at the university. Her mother, a proud Irish woman, actually convinces her to talk to directly with the NIH investigators Feder and Stewart. It is only then that O'Toole concedes, leading to the first true formal evaluation of the case.

Feder and Stewart receive the important 17 pages prepared by Reiss, review the article, and conclude that it was fraudulent. They write up a brief summary and pass it to chief Joseph Rall for his approval. However, Rall denies permission to print, which was particularly odd. Rall had been in touch with Baltimore, who undoubtedly influenced his opinion. Not to shy away from a fight, Feder and Stewart then take the case up with the *American Civil Liberties Union* (ACLU) who sent a note to Rall stating that his denial was illegal, as it constituted a case of 'prior restraint'— a blatant violation of the constitutional First Amendment. Only then does Rall concede and allows article to be published. However, when Feder and Stewart try to send the article to various important journals as *Cell, Nature, Science*, all journals

---

[234] These forced Feder and Stewart to consult renown civil rights lawyer Floyd Abrams, who noted that since the election of Reagan, cases as theirs tended to lose. However, Abrams wisely wrote *up New York Times* editorial, which in turn drew Mapplethorpe's attention to Feder and Stewart in the first place.

reject the paper. It appears that the accusation of fraud by a Nobel Prize winner was too atrocious an offense to merit any attention.

The NIH appears to initiate a thorough investigation of the case, but its first committee was composed of James Darnell and Frederick Alt. Darnell had been Baltimore's PhD advisor and mentor, and had also cowritten a textbook with his former student. Alt, in turn, had been Baltimore's student, cowriting some 14 papers with him. The members selected for the first committee represented an obvious conflict of interest, whose results would have been questioned immediately. Due to its glaring bias, the first review commission was scrapped.

A second NIH committee was then formed, constituted by more neutral separate parties: Joseph Davie, Senior Vice President of Research at Searle, Hugh McDevitt of the Microbiology Department at Stanford University, and Ursula Storb, a molecular geneticist at the University of Chicago. The second committee interview the relevant parties, gain access to the pertinent documentation, and look at the original graphs. While the second committee ultimately dismissed the case as one of fraud, it does request that a clarification be issued with regard to *Weaver et al*'s findings.

Dingell's legislative office also hears of Feder and Stewart's investigation, and views the O'Toole case as a David versus Goliath fight. Stockton, a member of Dingell Jr.'s staff, reviews the available evidence. When he presented the case to Dingell Jr. and the other members, all agree that it would be good case to pursue.

By this time, word of the affair was beginning to spread throughout the scientific community. When Leonore Herzchberg at Stanford hears the details of the case, she actually found O'Toole's arguments rather convincing. There were irregularities with the Bet1 reagent.[235]

---

[235] When she contacted Baltimore, who was by now receiving may letters about the gossip, he recognized that it was particularly important given their specialty and ranking within the community. Baltimore calls them directly over the phone and convinces the Hershbergs not to hire O'Toole. In short, he stains O'Toole's reputation before the Hershberg's eyes, though in the congressional hearings he claims never to have done so.

Becoming aware of the seriousness of the case, Baltimore proceeded to write 400 letters as a formal response to the allegations. In this, he was imitating Frances Collins, who had previously written 100 letters when it was found that his student's work turned out to be false. Collin's letters led to a particular set of papers to be retired, as these were no longer valid due to false data on which they had been based, and the case was generally regarded as exemplar.

Baltimore had previously depended on his personal lawyer when seeking counsel on the issue, as well as and MIT public affairs officers. However, upon the case's emergent escalation, he hires a Washington DC law firm with funds from his Whitehead Institute, suggesting the validity of the David-Goliath allegory. It looked like Dingell was going to have a 'trial in absentia' in that Baltimore was initially not called to be a witness at the first hearing.

The letter prepared with the Washington DC lawfirm's assistance characterized Dingell as a powerful figure that could greatly influence media and whom was undertaking a propaganda campaign within the scientific community. In his letter, Baltimore reassured scientists of the validity of the paper's findings, and argued that Dingell Jr.'s external intrusion into scientific affairs established a particularly troublesome precedent. Scientific debates could never be resolved in congressional hearings. Baltimore himself would take a legalistic position throughout the entire ordeal.

The Dingell Jr. committee investigation was long and drawn out. There were a total of five hearings which lasted a decade. The explicit purpose of Dingell Jr.'s hearings had been to supervise government funds. It was noted that a great deal of money was being handed over to the NIH, and it was its mission to determine how wisely such funds were actually being used.

The O'Toole case was so important, that even the Secret Service was called in to specifically verify the original documents. While the agency is typically associated jacketed armed men wearing earpieces and sunglasses to protect the US president, it also retains an extensive library on some 7,000+ ink samples which are routinely used to identify falsified currencies

232

floating in the market and gather substantiary evidence in important criminal court cases.

On multiple occasions, Secret Service officers went to Imanishi-Kari's laboratory to evaluate and take detailed samples. Incidentally, they did not want to know the meaning of its content, as they only sought to identify irregularities and changes in the physical documentation. When the laboratory's notebooks, images, and x-ray data were compiled, the US Secret Service found that Imanishi-Kari had written over many figures— particularly so when O'Toole raised the case before institutional bodies. Data that had allegedly been gathered posthumously in 1986 was identified as actually originating in 1984. All of the tape counters from lab had been typically advanced by 12 units per day. On one particular occasion, the tape had been shifted 1,534 times.

What was most shocking were the X-ray analysis, printed out in green and yellow ink, which were readily distinguishable colorings. The Secret Service discovered that the data for the final paper had actually been generated by Maplethorpe between 1981 and 1982, rather than the alleged 1984 date. The Secret Service's conclusions were irrefutable. The arguments in the paper could not have been possible, as transgenic mice had not yet been bred by that date.

The findings of the Secret Service wholly undermined Baltimore's ardent defense of Imanishi-Kari. At one point Baltimore even wrote a letter to Eisen at MIT, recognizing the fallibility of Imanishi-Kari's work and claimed that he would never work with her again. However, when Baltimore's letter to Eisen was leaked to the press on Sept. 8, 1986, Baltimore's reaction was to do a 180 degree turn and double down on his prior claims. The case appeared like it would go on forever.

There can be no doubt that Baltimore felt a strong sympathy for Imanishi-Kari, who held a press conference shortly thereafter. Imanishi-Kari's sister had died from lupus, and she herself suffered from the disease as well, although of a benignant type. This press conference, however, injured her case in the eyes of the scientific community. Walter Gilbert and Mark Ptashne were both leading scientists in field. Gilbert himself was also a Nobel

Prize winner. Ptashne, in turn, was a professor at Harvard who went down to DC to see for himself the hearings. Both concluded that the *Weaver et al* paper was false.

Gilbert came to the conclusion from viewing the Imanishi-Kari news conference. Personal feelings are never good motivations for scientific interest. For him, science needed to be dispassionate, using only cold guiding reason and evidence to come to conclusion, which would be otherwise distorted by an emotional personal motivation. When Ptashne read the *Cell* paper, he noticed it suggested that other papers had been validated. When he consulted the cited literature, he immediately noted the discrepancy—but in the public forum, the public had been left with the false impression that the paper had been scientifically correct.

Ptashne approached O'Toole during the hearings to let her know she was right, which stunned O'Toole. Her response to him, however, was urgent: she needed a job, which Ptashne initially took to be an emotional reaction. "All will be well.", he replied. But O'Toole responded: "I need a job NOW!" Ptashne served as consultant at the Genetics Institute, and was able to help O'Toole get out of an increasing financial morass.

O'Toole's voice strengthened as the hearings proceeded. At first she had been very nervous and tense, but as she began talking, she gained reassurance and confidence. It is quite odd that Maplethorpe, who had originally pushed for the escalation of the case, became the most cowardly of all during the hearing. Maplethorpe refused to testify, even when on the side of O'Toole, which forced Dingell Jr. to threaten Maplethorpe under the power of a subpoena.

Many scientists viewed O'Toole as the only one focused on the actual science of the paper, and its strict factual validity. Some had been publicly speaking without having even read the papers involved, and had taken David Baltimore's word 'for granted', providing ample evidence of the failings of collegiality in academia. Unfortunately this had also been the case for Steven Jay Gould, who mistakenly equated the affair with that of Galileo.

"Is there no challenge to a Nobel prizewinner?," asked Dingell Jr. during the hearings.

One of the most remarkable aspects of case was that the first NIH report actually congratulated O'Toole for her courage in bringing issue to the fore. However, upon publication and unbeknownst to its authors, this statement was mysteriously withdrawn from the final document. Dingell asked the NIH committee how this could have happened. Nobody knew. Senator Ron Lee Wyden of Oregon then questioned Weldon of the NIH. Weldon claimed that all fraud cases had been successfully resolved. Wyden proceeded to ask Weldon if he could name thee such cases—which he could not. When asked if he could name one single case, the answer was again that he could not. The inability to answer such a simple question clearly revealed that the NIH was not doing a thorough job of scientific overview. Some sort of substantive reform needed to be undertaken.

Pathetically, Baltimore had been able to get a large supportive crowds to attend the hearing, and even played the 'Jewish card' on the audience. Dingell Jr.'s aid Stockton had made some sort of allusion to the Nazis, which appeared in the *Boston Globe*. The Holocaust would not have happened if leaders had not looked the other way. As Edmund Burke once wrote, "The only thing necessary for the triumph of evil is for good men to do nothing." Baltimore noted he was a Jew and had been very offended by the comment.

But the fact of the matter is that Stockton's characterization was valid; Baltimore had simply not verified the work of Imanishi-Kari. Worse still, he had been using his enormous social stature to attack critics and squash all opposition—even those which represented a genuine scientific critique of the work.

This was one of the main points that Dingell Jr. focused on in his critique of MIT/NIH's reactions. The first institutional response seemed to suppose that a lowly postoc had to provide full evidence of wrongdoing before any charge of fraud would be evaluated. Dingell Jr. noted that such criteria would naturally prevent any whistleblower from revealing information in the first place, as they would be overwhelmed by its impossible requirements.[236] This critical point would be expanded upon by Paul Dotty, founder of the department of microbiology at Harvard. Dotty held great standing in the community.[237]

The final blow to the Baltimore case came when the NIH's Office of Scientific Integrity produced its final report. It had requested that Imanishi-Kari provide an index to all the original documentation, which Imanishi-Kari surprisingly refused to do. Her lawyer claimed that all contact with his client had to go through him, to which the NIH lawyer rebutted that Imanishi-Kari was directly responsible to the NIH as a recipient of its ample funding. It was estimated that the *Cell* paper cost some $82,000 to produce.

In spite of the fact that the NIH kept requesting information from Imanishi-Kari, she continually refused to collaborate with the investigation. Even newspapermen at the Imanishi news conference were puzzled by her behavior, as she was clearly injuring her own case. Imanishi-Kari was then warned that she would be placed on a probational list of fraud, and that all funding would be withheld until further clarification was obtained.[238]

Imanishi-Kari finally acceded to an interview in 1990, and her comments again were very odd. She claimed many of the problems stemmed from the fact that that did not speak English very well—a deceptive allegation at best. In fact, it was her use of scientific terminology that was hazy, specifically the use of the terms 'cloning', 'subcloning' and so forth. A clone of a clone would be called subclone, but then she would fail to differentiate between the two. She also claimed that she had a very good memory, and could remember every detail of her experiments— and could even recite details of a particular cell line. All the while, she had a very poor memory for dates, and rationalized the discrepancy by claiming that dates were not all that important. By

---

[236] The NIH essentially required the whistleblower to undertake its own complete and thorough investigation before looking into the allegation.

[237] He also had participated in the Manhattan project.

[238] In the hearing Dingell humorously asked whether the right hand and the left hand at the NIH ever spoke to each other. While Imanishi-Kari was being investigated by the agency, she continued to receive funds.

contrast, Baltimore well recognized the enormous significance of dates in a scientific record.

Imanishi-Kari's testimony was contradicted by that of Moema Reis, who was called via telephone to Brazil. Reiss claimed that she had to do all the work, and that it had been Imanishi who wrote down all of the data—including writing over some of the figures previously obtained. By contrast, Imanishi-Kari portrayed herself as being motherly to Reis, allowing her to develop her skills—which is doubtful, but in the end irrelevant.

Ultimately, Imanishi-Kari was found guilty of fraud, and was prevented from obtaining NIH funds. Annoyingly, the NIH final report caused even further reactions and retractions.

Baltimore was at first repentant upon its conclusions, and appeared to give an olive branch to O'Toole. By contrast, Imanishi-Kari's response in scientific journals was somewhat shocking, and continued to deny any wrongdoing or falsification of data. O'Toole's reaction was perhaps natural. In spite of the fact that she had warned Baltimore about the data, he had done nothing. O'Toole's public comments, however, led Baltimore to change his stance once again.

Paul Dotty's article was perhaps devastating, as it constituted a very harsh criticism of the scientific standards being set by David Baltimore. Dotty noted that Baltimore's actions sent the wrong message of a very low moral bar in the world of science; as if to say that when a journal published a faulty study, the authors had no responsibility whatsoever for retracting the original article upon discovering its errors. If you could get away with it, do so. This stance drastically lowered the standards of publication and the scientific enterprise as a whole. While Dotty recognized that there were many pressures in the academic world, he alluded to Richard Feynman's principle, first expressed at a Caltech commencement address.

Feynman argued that a scientist should account not only for what went right in an experiment, but also for what went wrong. One had to honestly deal with the bad data in the experiment, and attempt to confront it directly whatever that data might be. Science had high ethical standards, and its practitioners had to make absolutely sure of the claims that were being put forth in paper. If they did not, an external police agency that be would forced to oversee its activities—as had been occurring during the case.

A meeting was held at Harvard with MIT individuals, specifically between Eisen, Gilbert and Ptashne. Gilbert also emphasized the importance of Feynman principle, noting that self deception was the most common form of scientific error. We tend to look for evidence that validates our claims, without seriously taking into consideration evidence that does not fit our presumptions. It was also clear to him that too much collegiality dominated scientific relations. Scientist in short tended to trust each other too much, and were failing from taking basic steps to verify the information that was passed down the grapevine. In Puerto Rico is referred to as having too much '*amiguismo*'.

Baltimore exploded at Dotty's critique, issuing a reaction in the article "My dear Paul". The prior tone of repentance had now been completely upturned, and was again repeating ad nauseum that Imanishi-Kari's experiment was valid—a retraction which puzzled many scientist, but which ultimately cost him the Rockefeller presidency.[239]

In conclusion, it is interesting to note the parallels between Baltimore case and the story described in the movie *Spotlight*. In both instances, individuals belonging to socially powerful institutions do not want to reveal instances of wrongdoing, MIT and the Catholic Church respectively.

If it had not been for Dingell's investigation, it is likely that the case would not have received the attention it required. The NIH only brought greater attention to the case after Dingell Jr. initiated his investigation. Without it, O'Toole would never have gotten the recognition she deserved, as she was up against powerful actors who did not want their weaknesses exposed. She

---

[239] His position had been rushed forth at the institutional level, in spite of the fact that an informal poll showed that only 30% approved of his nomination. Rockefeller University was deeply involved in the world of biotechnology. The CEO of Bristol Myers was on its board of trustees. As the case escalated, the informal opposition increased to 75%. As mentioned before, leading scientists at the university resigned—the most shocking of which was Darnell's own resignation.

was already a mother of 3 kids, and would never have been able to continue in a career science were it otherwise.

How could we use sociobiology to analyze the case?

It is clear that reputation and 'long shadow of future' were prominently at stake, and as we have seen reputation influences reciprocal altruism. Higher status men will obtain greater benefits from their reputation than lower status men. We might suggest that as Baltimore's face appeared on the cover of the newspaper *The Boston Globe* next to the text of 'FRAUD', the cover likely overwhelmed him with fear of impact on his reputation. The measures he would take to sustain his reputation were of such a degree that they would ultimately come to undermine it. It is particularly striking that as more individuals became involved, these also sought to preserve their own reputations as well. If everyone repeats the lie, nobody will get caught. To this day both Baltimore and Imanishi-Kari refuse to recognize any wrongdoing, even though the most exhaustive OSI report clearly indicated so.

In the end, everyone was fed up with the case, which produced a lot of noise, and accounted for little scientific output of substantial value, to the point that scientific journals began refusing publication on the issue, as if beating a dead horse.

One should point out as well the enormous information plays in the long saga, which was a battle not fought on the battlefield but rather in the flow of information within a community. Today, online journals as *Retraction Watch* and *PubPeer* are critical means of scientific peer review. The OSI office was renamed Office of Research Integrity, which maintain an online list of the numerous cases of scientific fraud that continually appear. Strangely enough, while Baltimore's retraction appears on the webpage of the journal *Nature*, Paul Dotty's article oddly does not.

The saga continues . . .

Science and Democracy

What is the relationship between science and democracy, or between science and sovereignty? It stands to reason that if science is constituted by certain values whereby truth is established on the basis of evidence, it would imply a required dialogue between peers in any scientific discipline. Truth in science is not established by power but by reason. Other criteria acceptable in other social spheres are rejected in the realm of science—at least normatively. As such, the validity of a scientific argument is not based on the social traits of the proponent as nationality, race, gender, or religion, given that these are all irrelevant to the logic of scientific disputes.

In other words, the scientific ethos is associated with the democratic ethos. In democracies, politics is the result of deliberation between parties, which is to be contrasted to tyrannies determined by the role of power and as a consequence is arbitrary by nature. Autocrats as Joseph Stalin cannot know everything by definition of being mortal men. However, due to the enormous power they wield, they hold themselves as judges of everything about them, and as a consequence are inevitably bound to make mistakes—hence the arbitrary nature of rule is an inherent trait of all tyrannies.

As a result one would expect one social form, science, to be closely associated with the other, democracy. Science would tend to thrive in democracies, given the predominance of debate and dialogue in nonauthoritarian communities. Inversely, an increase in scientific activity of a collective would foment democratic political systems, as it would help establish ideal norms of a community. The willingness to question assumptions and claims would spread into public debate, and undermine arbitrary decision making processes of tyrannical political leaders. Historical examples do suggest that such relationship indeed exists.

Many of the United States Founding Fathers had a strong interest in science, some of which are considered to be members of the Enlightenment, as Thomas Jefferson and Benjamin Franklin. Both men had scientific achievements and both obviously participated in American Revolution–which represented a real threat to their lives. If unsuccessful, these could have been executed. Their biographies suggest that their scientific and political activities were intimately intertwined.

Puerto Rico is another such case. The end of the nineteenth century is generally regarded as a cultural golden age in the island, dominated by highly educated men as Roman Baldorioty de Castro, Jose Julian Acosta, and Ramon Emeterio Betances. All studied science in Europe, the first two in Spain, and the third one in France. It is particularly striking that many of these pushed for insular independence, particularly Baldorioty and Betances. We may also point out that while Acosta was a moderate, he was an active abolitionist in Puerto Rico.

While the political and scientific achievements of Puerto Rico's 'founding fathers' were not of the same degree when contrasted to their US counterparts, the association between the two activities is correlated.[240] In both cases, scientists sought out political reform so as to end colonial regimes. Their ambitions included greater autonomy for their respective geographical regions, as well as advance local scientific achievement. Yet, how does the scientific ethos exactly influence political structures? Is there a broader more generalisable universal pattern? In other words, what is the relationship between science and freedom?

Similar historical debates have been held regarding the relationship between the Enlightenment and the French Revolution. While the Enlightenment occurred in many places, it was a particularly 'French' phenomena, profoundly influenced by the Scientific Revolution. Men as Voltaire sought to diffuse its

---

[240] It might be said that politics undermined their scientific activity.

findings and values throughout society, and produced translations of Newton's work. Voltaire distinctly used it in his war against the Catholic Church, but it was not the only institution attacked. Tired of the abuses and prestige of monarchy, the French Revolution beginning in 1789 produced the *Declaration of the Rights of Man*, establishing universal suffrage as a key human right.

Did the Scientific Revolution ultimately lead to the French Revolution? The issue has been debated since its emergence more than two centuries ago. Studies done by I. B. Cohen on the phraseology are suggestive but not definitive.[241] The issue is a complicated one, and has sparked many debates with regard to the political effect of science; as its relation to democracy, sovereignty, and nationalism. Writings on nationalism have grown tenfold, while volumes dedicated to the issue in the Library of Congress have increased forty-fold. The topic is useful for what it reveals about the character of science and its social implications; a number of insightful generalizations have been produced.[242]

Contrary to popular opinion, Robert Merton did not claim that science inevitably led to democracy, noting that science has occurred in many diverse settings. The establishment of the Paris Academy of Science occurred in France under Louis XIV. Charles II in England set up the Royal Society of London, while Peter the Great in Russia created the St. Petersburg Academy of Science so that Russians would not be referred to as barbarians by the rest of Europe. The aforementioned cases were obviously monarchical political systems where the sole point of reference and truth was the king.

---

[241] Early on, the term 'revolution' did not have political implications, and referred only to celestial movements involving a profound change of view. Over time, there was a gradual political adoption of term to represent how it is used today: radical institutional change.

[242] Many have written about topic, including Robert Merton, Edward Shils, Ernest Gellner, David Thompson, and Don K. Price. The first to have specifically touched on the issue of science and democracy was the historian of medicine Henry Seigrist in 1938.

That being said, Merton noted that scientific activity is based on a particular set of values and mores which may or may not coincide with its social setting, thus placing it either in conflict or harmony with its political system. Undoubtedly, science has been incongruous with many social settings and contexts, and has even tended to conflict with nationalism. While in theory all talent is open to science such that anyone with ability could enter its realm, at the onset of WWI German scientists produced a public letter glorifying German science over that of all other nations. In this case, the scientists had written as politicians, violating well established scientific norms. Max Planck later came to regret signing it.

Science has also tended to conflict with social settings such as capitalism and industrialism, where the tendency to appropriate knowledge directly clashes with its universalist traits. As knowledge is held to be a common good, there is the trait of 'communalism' in its practice. The main prize of scientific discovery is recognition, hence why an eponym is also commemorative. The term "Boyle's law" both remembers and celebrates Robert Boyle long after his death.

Merton did note, however, that certain social settings were more favorable to scientific activity because the prevailing social values coincided with those of science—and in particular those of democracies. While many tyrannies have had science and scientific activity, there is a naturally inhibitory factor which limited its success, as the values between the two naturally clash and science undermines the particular political system in which it is located.

David Thompson has noted that the two most profound changes which redefined role of governments were the American Revolution (1776) and the French Revolution (1789). The proposed norm that political legitimacy is the result of the consent of the governed was groundbreaking; all had equal rights to political power. By contrast, tyrannies traditionally sought to evade the consequences of universal suffrage. Voltaire, Montesquieu and Rousseau sought a profound reorientation of society.

The emerging paradigm of the era became the universality of rationality; science was not the result of genius, but accessible to all who were interested in exploring it. This idea was pushed by Bernard Le Bovier de Fontenelle in his obituaries, where he spread specific values of in his funeral orations. The universality of rationality at the time implied the universality of conclusions, with profound implications for democratic political systems. If all knew science, all would arrive at same conclusion, in theory; science implied that universal consensus was possible in a democratic community.

Increasingly, Enlightenment thinkers came to regard education as critical for a functional democracy. Structural political modifications of despotic rule were ultimately futile, and these ideological changes served to undermine monarchical rule. The sole criterion of truth was no longer the king, as the well known case of Louis XIV.[243] As a reaction to these challenges, monarchs of the period began using the term 'great' to bolster their status, as the case of Catherine in Russia or Frederick in Prussia. In a shifting ideological terrain, monarchs witnessed the foundation of their political legitimacy severely eroded. Scientific and rationality undid the cultural basis of their political power, so much so that their attempted reforms to ameliorate increasing public dissatisfaction were not generally held as credible.

Marie Jean de Condorcet and Antoine Lavoisier strongly pushed for educational reform. Tragically, both witness and approved of the emerging French Revolution, whose chaos could be attributed in part to the low state of education at the time in France. Debates during French Revolution were utterly chaotic.[244]

There was a tendency to detest organizations; all institutions tended to be ruinous of reason in that these often seek to make

---

[243] As the 'God king' who was publicly deemed to represent God on Earth, all others had to unquestioningly comply with his rule.

[244] Tragically both were killed during the period. Condorcet had briefly headed the Academy des Sciences of Paris , and began the first true statistical analysis of society, which was revolutionary for its time. Lavoisier in turn had just revolutionized chemistry. For him, the use of appropriate names and technical language would lead to new discoveries in the field.

themselves exempt from public norms. While some obvious cases included the Catholic Church, there were many other examples, as the guilds of medieval Europe, and political parties which undermined the rationality of democratic politics. The use of the term *political parties* was purposefully removed from the US Constitution for these reasons. Thomas Jefferson is known to have said that if he had a choice between ascending to heaven with a party or going to hell, he would choose the latter. George Washington himself banned all talk of political parties in his farewell address. It was generally hoped that the incorporation of science into politics would eliminate need for political parties; common rationality would lead to common solutions. Other historians and political theorists have taken similar positions.[245]

---

[245] Edward Shils has argued that the role of politics in a democracy was truth-seeking. Political activity was not reducible to principles but was rather a constant ongoing process. There is not a 'core' to social life, as in the peeling of onion, but rather society was made up of a web of relations in constant flux. As such, politics was characterized by a tension between potentiality and actuality; political activity seeks utopia which are never achievable.

Due to the never-ending existence of problems requiring solutions, and the need for a constant ongoing debate, civility in liberal democracies became an essential component. As each political actor has interests which will inevitably conflict at some point or other, civility was required in a rational dialogue for their resolution. The inability to reach a compromise would completely stall the political process; by undermining some of one's own interests, all benefited as a result of dialogue. In short, civility helped insure rational exchange of ideas. Civility was the transfer of scientific culture onto a political setting. Implicit in this arrangement was also the notion that certain irresolvable topics would not be discussed, as it would lead to an irrational debate guided by emotion and passion—religion being one such topic. Rational discourse is impossible on matters of faith.

Ernest Gellner for his part noted the effects of science in modern setting, making a distinction between the intelligentsia and the intellectual class. The former are newly educated minority, who have typically studied abroad and thusly exposed to a different set of ideas. Although the intelligentsia speaks the same language, they have drastically different cognitive models, and to a degree are 'alienated' from their own societies. They are a part of but do not necessarily belong. By contrast, the intellectual class comes from the society that produced it, sharing the same values and presumptions as the rest of their community, and consequently seek to reproduce this value structure institutionally.

Don K. Price's *The scientific estate* (1965) was radical for its time, upturning common notions which perhaps had been too idealistic. In contrast to Merton or Thompson, Price is much more cynical of the role of science in society, imbuing it with shades of tyranny. The prior major threat to freedom had been the Inquisition, but in his work Price viewed science as an even greater menace. Situating his work in the modern setting of the Cold War helps account for such harsh views.

For Price, politicians generally lacked the capacity to understand and fix modern problems, whose nature was simply too complex due to their scientific origins; issues as pollution required a level of sophisticated analysis that was orders of magnitude beyond the political norm, which in turn led to a structural dependency upon scientists by politicians. Scientists were ironically needed to fix the very problems they had created. Eisenhower's comment regarding the "military industrial complex" is an allusion to the process whereby thegovernment became a captive of the scientific elite.[246]

He also points out that scientists were not neutral actors. Historically they have been rather naïve, believing they could easily 'fix any problem' if enough attention was given to the issue. A scientist's close affiliations to a particular corporation or university, however, biased his advice. One of Price's

---

The role of science is therefore 'revolutionary', particularly so for very traditional societies as Turkey, India, Brazil. The scientific intelligentsia force changes in attitudes and practices to those which more consistent with the values of liberal democracies and of science generally speaking. These, for example, push for meritocracy in labor selection, or the criteria of employment on the basis of ability rather than family connections. Attempts at nationalization under Gellner's schema is thus often a reversal of the modernist values embodied by the scientific intelligentsia, in that such institutional changes seek to reinforce traditional class values as well as prevent the intrusion of external agents in economy by creating protected economic areas. Nationalist policies is thus seen by Gellner as a reduction in the role of meritocracy via the abandonment of open competition.

[246] Price depicts the radically different psychological modes of being. While the scientist typically asks 'is it true?', the politician wonder why his opponent is taking the particular stance they are. In the competitive arena of politics, actors were constantly jousting for position, truth becomes only a weapon of civil competition.

contributions was to depict the close integration between the two. Universities as MIT, Caltech, University of Chicago, and Johns Hopkins received 3/5 to 5/6 of their total budgets from federal funds, and thus sought to enhance governmental dependency via consulting.

The role of industry was particularly noxious, in that contracts established for projects were no longer decided on a competitive bidding process, whereby the lowest price was selected, but were now open ended. Project estimates could never be determined as these constantly fluctuated , which was particularly noxious to governmental fiscal planning. While industry tended to establish strategic outlines in 10 to 20 year periods, politics was ruled by 4 years cycles, producing two mismatching cycles which naturally clashed.

Ironically, corporations did not actually contribute that much to science per se, but rather did so only under particular settings. These might include monopoly conditions, as the Bell Telephone Co., or giants with wide range of products as DuPont, which in both cases invested heavily on research. In spite of this, government typically paid for 3/5 of all research and development. Worse still, the United Sates now found itself in continual preparation for war, which had previously only been the exception, as in the case of WWII.

The end result was that science degraded democratic institutions.

The modern political interpretation of science could be sharply contrasted with its origins. It was believed that the Newtonian notion of counterbalancing forces establishing harmony in nature could also be applied to establish harmony in the world of politics. Both Thomas Jefferson and Adam Smith believed that the government should have no participation in the economy, and the ideal politician was one who did not have conflicts of interest which would biased political decision-making. By contrast, in Price's era, there was a tight integration between the two. The Speaker of the US House of Representatives, Tip O'Neil of Texas, did not meet NASA officials of the when they visited Washington DC, so as to send a clear message that he wanted their facilities to be placed in

Texas. Research centers attract industry, which in turn created jobs and political power.[247]

In our brief review of political theorists, the positions regarding the relations between science and democracy are mixed, and there is no current consensus in spite of the insightfulness of the diverse positions. Perhaps a few historical cases will help clarify the issue, specifically those of the United States and Puerto Rico.

Benjamin Franklin was one of the oldest signatories of the US Constitution. He also recognized Jefferson's impressive writing abilities as a young man, and convinced him to write the Declaration of Independence.[248] Franklin is also known for the myth of flying a kite in the middle of a thunderstorm. While the widely depicted image somewhat distorts his scientific contribution, it does reveal Franklin's precocious scientific inquiry.

To what degree were the two related? Is there a relationship between his search for US independence and his scientific ethos; which is the cause and which is the effect?

Franklin's work in electricity is much more important than is generally recognized. Prior to him, the field was in chaos with a great many nondefinitive experiments, some of which were humorous, but which lacked a comprehensive theory.[249] Franklin's work in the field was so important, that he was referred to as Newton by Joseph Priestly in England. The colonial scientist had provided first comprehensive theory of electricity as well as postulated the notion of conversation of charge. To understand how insightful his experimental work was, J.J. Thompson, discoverer of electron, said that he tended to look at Franklin's writings to help understand electrical phenomenon under study or to suggest new experiments.

---

[247] Tip O'Neil was one of the longest serving congressmen in the US legislature.
[248] However, it was ultimately a draft reviewed and revised by Franklin and John Adams.
[249] In some, a boy would be held in mid air and 'cracked' with static electricity.

Franklin's key observation was that electricity consisted of a single substance, a fluid of sorts, that was passed between objects. Lighting as such was electricity as well, in that a charge was built up as clouds moved. With a kite, Franklin be able to access charge in the sky and fill up his Leyden jar, the battery of the time, thus proving lighting was electricity. Franklin also showed that the charge in the Leyden jar was not accumulated in the water or in the metal plates of jar but rather on the glass which held them together, which led him to cover the glass with metal forming the basis of today's modern car battery.

His work also led to the innovation of the lighting pole, which transferred the charge from a lightning bolt directly to the ground, bypassing the building. As most colonial structures were made of wood, and the rate of lighting discharge to the ground is inversely proportional to its distance from the ground, the rate of death and destruction were fairly high in the colonial period. In early 18th cent Germany, some 386 churches had been destroyed; in Venice Italy some 3,000 died when lighting struck a church used to store gunpowder. Given these accomplishment, it is no wonder that Franklin was referred to by Immanuel Kant as the "new Prometheus". His countless innovations improved the quality of daily life worldwide.

It would be ideal to suggest that Franklins' science influenced his politics, but his biography reveals that relationship is hard to trace. It is certainly the case that independent mindedness ran in family; 'Franklin' is a derivative of 'free thinker', and had been chosen by family as its last name to embody these values. Franklin was only in school until 12 years old.[250] However, while did not receive much of an education, he loved to read, and was essentially a self-taught man who pushed himself to his limits.[251]

He was forced onto an apprenticeship with his brother, who had a printing business, and produced the first independent paper, the *Courant*. Franklin learned a great deal while working there, in particular how to criticize without personally attacking another.

---

[250] Franklin's scientific abilities were limited because of his poor education; his mathematical training was very shallow, and thus could not push the science of electricity very far.

[251] His *Autobiography* is an entertaining and insightful read.

A key value embodied in his works was that of civility. He actually wrote under various pseudonyms, at one point pretending to write as an old woman; everyone was surprised when the true author was revealed. Franklin was deeply empathetic, allowing him to enter an individual's worldview and mentality.

He was also an active promoter of public institutions in the United States. When Pennsylvania's government refused to act after the French-Indian incursions into Philadelphia, Franklin formed a militia to protect its citizens. He also formed the *Junto*, an informal association of intellectuals whose single criteria for membership was to have observed British encroachment of civil liberties. He is instrumental in the creation of libraries to spread useful knowledge. He began the first volunteer firefighting department, an institution we take for granted today.

Without a doubt, Franklin was institutionally creative, forming all sorts of social groupings for particular tasks. Alexis de Tocqueville observed the contradictory nature in US society between individualist and communal activity, which for Franklin represented no contradiction whatsoever as the two went hand in hand.

What then, was the relationship between Franklin's science and his political activity? Did science lead to his political search for US independence, or was it the other way around? It appears as if the two coexisted, mutually feeding off one another. Franklin used his science for political purposes, given the fame he had achieved in his own right. He had been elected to the most important scientific society of the era, including both the Royal Society and the *Academie des Science* (Paris), as well becoming recipient of the prestigious Copley medal.

Franklin not only relished his scientific reputation, but was careful to cull his public persona as well.[252] He wisely used his scientific fame for political ends, using it to convince the French

---

[252] While in Europe, he used a distinctive "Davey Crockett" raccoon hat to characterize himself as a humble woodsman. Tales as the leaving of candles near windows late into the night to suggest continuous work activity are found throughout his autobiography.

to support the independence of a future United States—which provided critical military and political support to the American cause when it was most needed. While Franklin's case is suggestive, its is unlikely that science directly led to his political activism. His biography suggest that both were part of a common cultural heritage and mindset.

Thomas Jefferson was a very different character to Franklin. The latter had a highly amicable political personality and never actually expressed his own personal opinion and preference.[253] By contrast, Jefferson was a much more private individual, who tended to be sought after for political candidacies rather than actively seeking these out. When George Washington wins the presidency, he insisted that Jefferson become his Secretary of State—a position Jefferson was very hesitant to take. Unlike Franklin, Jefferson was never as popular a figure.

There was another key difference between the two: Jefferson had been very well educated. His father had been a surveyor, whose mathematician friend provided exclusive lessons. Jefferson's' scientific and technological achievements stand alongside those of Teddy Roosevelt's, the only scientist to have become president. Both were avid natural historians, but Jefferson so far is the only president to have read Newton—and understood him. Jefferson even used Newton's fluxion (calculus) to design the most efficient plow structure.

While living in France, he was impressed with its intellectual world, but shocked by its social one. He came across Lavoisier's work, but did not believe it as man's senses were too limited to come to any conclusions about chemical elements. He also noticed that men had taken over women's professions as that of cooking, and hypothesized this to be the reason for the high incidence of prostitution on its streets.

In light of this, it is somewhat striking that Franklin's scientific achievements far superceded those of Jefferson's, whose greatest formal contributions were technological rather than scientific. Nonetheless, his *Notes on the State of Virginia*

---

[253] Adams was routinely infuriated by Franklin, who could never be entirely sure of Franklin's actual position on a given subject.

was a remarkable achievement for its time. Jefferson not only established the science of stratigraphy but, offers a rebuttal against Comte de Buffon's claims regarding the degeneration of species and humans in the new world. The debate would have far reaching political and economic implications.

Georges-Louis Leclerc had been in charge of the *Jardin du Roi*. A mathematician by training, Buffon then began comparing species, and offered fascinating ideas in his *Epochs of Nature*. He argued that the Earth was much older than claimed in the Bible— a conclusion he reached by studying the cooling of melted iron. His approximation of 75,000 years went directly against religious claims. Similarly, because the Earth cooled, time ultimately had an arrow, in that distinctive historical change was irreversible. Buffon brilliantly proposed the notion of periodic animal changes, tying biological changes to geological ones. As the Earth irreversibly cooled, new species were continuously being formed at the poles, which were gradually forced to migrate to the tropics.

While these views now appear to be somewhat primitive from our point of view, Buffon was on the right track. Unfortunately, he also claimed that new world was inferior as well. Its animals were smaller, weaker, and duller; its human types were less manly, duller, and had little vitality as noted by the lower rates of reproduction. To top it off, animals and humans that traveled to the new world also degenerated in mind, body, and spirit.

It was obvious to Jefferson that Buffon had not carefully researched or thought out these arguments. At a dinner with Buffon in France, Jefferson asked the Americans at the dinner table to stand up, as well as the French delegates. It was clear that, on average, the US delegates were far taller and larger than their French counterparts. As it was a small sample, Buffon did not accept the refutation.

Buffon's claim, however, was no trivial matter. If accepted, it had enormous political implications. As the United States was a young nation, it could not afford to be so categorized, being detrimental to its commerce and international relations. In spite of their many internal divisions, specifically the Federalists anti-Federalists bands, all gathered and provided data to Jefferson in a defense against its outrageous claims. Even Alexander Hamilton, Jefferson's nemesis in the Washington administration, helped Jefferson out with the task.

In contrast to Buffon, who was clearly extemporizing, Jefferson undertook a systematic tabulation of animals between the two continents. Of the 18 species unique to Europe, Jefferson found that their average weight was 27 lbs, or half that of the 27 species unique to the United States. The sheer number of bird species was remarkable; Virginia alone had 170 bird species. It was clear to Jefferson that Buffon had been careless in reaching his conclusions, and would have to retract them. Unfortunately, Buffon died before such a retraction is ever issued.

While the Buffon-Jefferson debate might be considered as a case for the influence of science in politics, it is so in a negative sense however. Did science have a direct influence on US politics? I. B. Cohen is perhaps the only historian of science to have most seriously studied the issue. He noted that the analysis of the US constitution is typically framed within political history, specifically the philosophical notions of John Locke. Because Cohen was a historian of science, authoring important works on the Scientific Revolution and editing Newton's *Principia*, he was able to look at underlying scientific influence underpinning its foundation.[254]

Cohen argues that Newton's influence is distinct in Jefferson's political writings, such as the borrowing of its language. In the Constitution, we find phrases as 'natural law' and 'self evident truths' which came directly from Newton's axioms. In contrast to the truths of natural history, mathematical truths are self evident and unquestionably established using Euclidean logic.[255] Thus when Jefferson writes "We hold these truths to be self evident", he is implicitly referring to irrefutable truths akin to those in geometry. The logic behind the notion that men are endowed with certain inalienable rights as life, liberty,

---

[254] Cohen's *Birth of a New Physics* was the standard textbook in the history of science for many years.

[255] The biological truths of natural history, on the other hand, merely rested on observation, and hence were potentially subject to change with new data.

pursuit of happiness was one fundamentally drawn from the world of science, according to Cohen.

More importantly perhaps, Jefferson sought to endow US institutions with the principles of rationality and civility, whereby its protocols and procedures would encourage men to dialogue about issues reasonably, rather than by appealing to their lower instincts. Jefferson sought to intimately weave Enlightenment values of science into the very fabric of American government and society.

We may cite multiple examples. As Secretary of State and head of the Senate, Jefferson introduced procedural guides which now form the basis of Senate and House rules. Specifically, bills could not be introduced by the most popular legislators, but rather had to be introduced in a specific logical order. The process thereby provided all senators an equal opportunity to introduce their measures rather than giving preference to the most powerful ones.

Jefferson also attacked the unequal distribution of representation that had been first introduced by his nemesis, Alexander Hamilton, which would have favored geographically smaller states over larger ones. While the determination of the number of delegates per state might appear to be a simple issue, the problem is more difficult. On deeper inspection, the Hamiltonian plan proposed the use 'simple ratios' or fractions. Hypothetically, if Connecticut had 2.8 senators and Virginia 3.1, the these would be rounded down to 3 in both cases. However, simple ratios tended to favor smaller northern states against large southern states, which Jefferson represented, and did not provide for a consistent division. By using the largest state as a common denominator, Jefferson's plan produced a more uniform distribution. Washington agreed, and revoked the Hamiltonian proposal.

Finally, Jefferson was instrumental behind the establishment of a Bill of Rights. One of the key elements to any rational government is that it cannot be allowed to act arbitrarily, but must rather must operate within certain norms and strict parameters. Jefferson saw the Bill of Right as a fundamental measure which placed clear and distinct limitations on state power. Jefferson equally believed in the freedom of religion, in that relationship to God was between the individual and his Maker only. In light of the irrationality of religious debates, there also had to be a strict separation between the two.

Can we then argue that science had a political influence in Jefferson? While it is difficult to suggest that science pushed Jefferson to the adoption of independence, there is no doubt that it helped to insure a more democratic government. If tyrannies by nature tend to be arbitrary, then just laws helped to insure the stability and longevity of America's political experiment in democracy. The government, as all natural law, had to be applied equally to all, regardless of their station in life. Jefferson the Virginian called for the smallest federal government, in contrast to Hamilton, his Caribbean-born counterpart.[256]

The cases of Franklin and Jefferson are tantalizing. There can be no doubt about their scientific achievement and ability, as well their leadership roles in the formation of a new sovering nation. Equally, it is clear that science influenced their political activity. Franklin used his scientific popularity to political ends, while Jefferson imbued US institutions with the rule of law. The shadow of science is very much present, and strongly affected the character of their political behavior and actions.

However, to claim that science somehow established US independence or sovereignty, is a difficult argument to accept. Could the same thing be said of Puerto Rico?

Puerto Rico's version of the US eighteenth century Enlightenment occurred during the nineteenth century. The explosion of a vibrant intellectual and political community implied an unexpected and unforeseen critical mass of talent. The latter half of the century is truly one of its most impressive periods, resulting in the formation of the Puerto Rican identity. We might refer to it as the '*generacion de los proceres*'.

There were many individuals associated with Puerto Rico's late nineteenth century golden age, but we will focus only on

---

[256] Jefferson was a Virginian who called for a small federal government, and viewed the US as an agricultural nation. Hamilton for his side defended banks and was curiously from the British Caribbean, having been born in Nevis and raised in St. Croix.

three, whom happened to be most scientific: Ramon Emeterio Betances, a physician seeking independence, Roman Baldorioty de Castro, a botanist-physicist seeking autonomy, and Jose Julian Acosta, a printer who sought to end slavery in the island.

It is somewhat hard to classify Ramon Emeterio Betances as a nationalist. His mother dies when he is 10 years old, whereby he is 'exiled' to France in 1837. His father's friend, the pharmacist Jacques Maurice Prevost, promises to takes care of the young Betances during his long formative period in France. Betances naturally becomes fluent in French, turning into a type of mother tongue. He participates in the 'Second French Revolution', where thousands took to the streets, in what must have been a profoundly influential experience.

He then studies medicine at the University of Paris (Sorbonne) with leading scientists as Paul Broca. Being closest to his younger sister Demetria, she becomes jealous as it became clear that her own education had not been as high of a quality. When Santo Domingo explodes in a revolution in 1861, Betances travels there, spending nearly every penny he had. His role became so prominent in the neighbor island, that its own Captain General orders his exile in 1864.

'*Antillanismo*' became Betances's common motif, alluding to the goal of political freedom throughout the Hispanic Caribbean. The principle was not just political freedom from either the Spain or the United States, but included abolitionism, as well.[257]

The prior data suggests that Betances's variant of sovereignty has to be circumscribed. While we can obviously say he was a nationalist, his appeal seems to have been more of a principled and generalized call for independence throughout the Caribbean. Sovereignty seems to have been more of a 'philosophical' notion than an emotional one based on gut patriotic feeling of a particular nation state. His departure from Puerto Rico at such an early age must have been very traumatic, given the context in

---

[257] It goes without saying that Betances was of African American heritage.

which these incidents unfold. Upon his own father's death in 1847, he returns to Puerto Rico when he meets his bride to be.[258]

To what degree could we claim Betances was Puerto Rican? This is not necessarily an easy question to answer.[259]

We might point out that while in France he did actively seek out other Puerto Ricans, and was friends with Francisco Oller—a Puerto Rican who formed part of the French Impressionist movement in art. Betances also organized with Acosta, Baldorioty, Tapia, and others the *Sociedad Recolectora de Documentos Históricos para Puerto Rico*: a larger group which compiled Puerto Rican documents found in Spanish archives, as very little was known about the history of the island within the island itself. Their findings were edited and published by Tapia in the *Biblioteca Histórica de Puerto Rico* (1854). Other members of the group included Segundo Ruiz Belvis, a lawyer who would help him on occasions. Betances clearly kept abreast of islanders in Europe, suggesting the deep emotional bonds of his personal nationalistic sentiment.

Betances's scientific achievements have not been well studied. He is well known for his participation following the cholera outbreak of 1856, where 25,820 died—of whom 80% had been *pardos* or free blacks. Betances joins a commission six doctors to control cholera's spread in the western portion of the island. The exact cause and factors were unknown, but Betances successfully establishes a quarantine which helps curb the epidemic's growth and impact.[260]

It is worthy of note that while in Paris, Betances had coauthored works with a number of important scientists from the period, including Geoffrey Saint Hillarie of the Pais *Museum of Natural History*. In spite of his revolutionary activity, Betances

---

[258] Betances had recently returned from France, where he had married his niece, whom suddenly dies of Typhoid fever. The guilt and trauma of the incident lead to the novel *Virgen de Borinquen*, published in the same year as Darwin's *Origin of Species* (1859). It is her death which puts him back in Puerto Rico, and the island symbolically becomes his new 'wife'.

[259] His grandfather had migrated to Puerto Rico as a result of Haitian Revolution.

[260] All infected patients were quarantined for up to a month, as were newly arrived ships.

continued his scientific research throughout the rest of his life. In 1886 he published a study on the *uretrotomía* arising from sexually transmitted diseases. In 1891, a French scientific journal cited two of his works, one of these being a study on tuberculosis. He had also written on elephantiasis (filariasis)—a striking fact as it had been this disease which led to creation of parasitology and tropical medicine by Patrick Mason in Hong Kong between 1866 and 1889.[261]

Whatever Betances's scientific contributions might have been, it is pretty clear nonetheless that his productive life was focused on political activism, seeking social changes as independence and abolition. He is known to have purchased the freedom of many Afro Puerto Rican infants, buying these at 25 cents to set them free. His case is perhaps similar to that of Sun Yat Sen, a young Chinese student who studied medicine with Patrick Mason. Rather than enter medicine and biological research, Sun Yat Sen became a revolutionary leader in China, and later president of the Southern Chinese Republic. Assessing the life and legacy Betances is difficult however, given the absence of a comprehensive archive of his writings.[262]

Ramon Baldorioty de Castro can be seen as a counterpart to Betances. Betances had asked Baldorioty to participate in the 1868 Lares Revolution, which the latter declined, believing in the peaceful means to political change. Baldorioty tried to operate within the political system to reform it from within, and wrote many proposals and reforms to this end. Of all the intellectuals of the period, Baldorioty can be seen as the closest to a Jeffersonian type who sought to rationalize Puerto Rican society and government.

---

[261] The particular cases treated by Betances shocking, as those of testicular tumors of up to 26 lbs, which were obviously debilitating and humiliating. Apparently Betances did create successful 'treatments', helping to ease the pressure and affording some functionality to the patient. In one case, Betances had leather pants made, to control swelling.

[262] As noted in the film "*El Antillano*", Betances's historical memory has been erased—as when 100 notebooks full of personal letters disappeared from a Cuban archive.

Some of Baldorioty's proposals sought to overturn these injustices with the creation of a formal banking system. For example, as there was no currency given the absence of a banking industry, merchants served as high interest money lenders—a historical problem in Latin America. As Jefferson, Baldorioty also wrote a critique of Buffon's degeneration theories, particularly with regard to local laborers. It was claimed that the local Puerto Rican was inherently lazy. Baldorioty pointed out that, if this was the case, the *libreta* system used to regulate and control his labor would have been useless. Baldorioty's own study of the *libreta* system actually revealed that it led to a reduced productivity, a finding which contradicted governmental claims. If 'facts' served as justifications for unjust political systems, then careful evaluation of these could help undermine unjust and corrupt policies as well as the governments which promoted them.

Baldorioty's moderate stance can be said to be a reflection of the island's political circumstances of the period. Although a Spanish colony, the political orientation of the diverse military governors varied greatly. At times liberal governors held power, and pushed for educational reforms as the creation of schools. These, however, would alternate with their conservative counterparts, who then sought to undo liberal reforms. The radically divergent stances of the respective governors implied unstable social policies and undermined all public institutions. Some governors might even shift at midterm, arriving as liberal dictators only to become conservative demagogues by their departure.

There were deep structural problems as well. Military governors were dictatorial emperors in the island, with control over all three branches of government, violating that cardinal separation of powers which stands at the core of the US constitutional system. The military governors also appointed *alcaldes* (mayors) to their municipalities (*'municipios'*), thereby keeping a firm grip over the entire island.

Baldorioty's moderate stance might have been unwarranted. The military governor Laureano Sanz routinely persecuted him. Often such persecutions tended to be mere hassles; the mortgage

259

Baldorioty to obtain his house was unexpectedly withdrawn. His salary for teaching at school often went unpaid for months at a time, and so forth.

This persecution likely pushed him into a more liberal stance of autonomism. Baldorioty joins the liberal party, giving it a new lease on life. As a man of science, he could see more clearly than others. Tragically, however, during the *Compontes* of 1880 Baldorioty is jailed in *El Morro*, which becomes his death sentence. He would have died there had it not been for the pharmacist Arriaga, who goes to Spain to complain of Sanz's abuses of power. This brave act thus led to the removal of the powerful military governor, and to Baldorioty's consequent release. But his mistreatment at *El Morro* had ruined his health, and he dies shortly thereafter. The city which had idolized Baldorioty later crucifies him.

Baldorioty's life might be characterized as tragic. He was constantly under the threat of poverty, and at one point has to ask a friend to provide food for his family. More tragically still, there can be no doubt of his intellectual merits fell onto the deaf ears of the local authorities. His attempts to form a leading educational institution in the island results in a needless and grotesque public humiliation. Like so many Sir Galahads who wasted their life in the tropics, as Manson noted in China, Baldorioty's efforts at scientific and political reform were routinely blocked by Spanish colonial authorities.

Baldorioty's scientific credentials were impressive indeed. In spite of his mother's meager wages, his remarkable intelligence is noticed and he is accepted into Rafael Cordero's humble primary school, where many *proceres* were originally educated.[263] At the primary school, he makes many early friends, as Alejandro Tapia y Rivera and Acosta. Both he and Acosta enroll at the *Seminario Conciliar*, where they obtain their first scientific education from Padre Rufo.

Padre Rufo is able to obtain funds for a group of distinguished students to study in Spain. Of the four students,

---

[263] Due to his father's mistreatment of his mother, Baldorioty as an adult changes his original last name from Castro to that of his mother's (Baldorioty).

Acosta and Baldorioty were only supposed to study pedagogy, while the other two fellow classmates were to study science. However, three of them get smallpox, but since Baldorioty was already immune, he tends for the others. Only Acosta survives, leading Rufo to shift the original funds of the *Sociedad de Farmacéutica* to the second pair. Pharmacists at the *Bótica Babel* served as early scientific rebels in the island, akin to Franklin and Jefferson.[264] Padre Rufo, a scientific exile from Spain, also provides funds out of his own pocket for Baldorioty's European higher education.

Baldorioty became well versed in calculus as well as in botany, which becomes one of his first teaching jobs on his arrival to the island. His intellectual abilities are well recognized, in that both he and Acosta were disciplined students who recognized the importance of education for the development and wellbeing small island.

When inquiring into Baldorioty's scientific achievements, we come across a common pattern in the local proto scientific-intelligentsia of the time. After an early period of scientific activity, much of their focus then turns to political activism. The cases suggest that despotic environments hostile to scientific activity force scientists to shift course and enter into politics and political activity. Again, this had also been the experience of Sun Yat Sen in nineteenth century China, suggesting that Merton's claims might have to be modified.

Much of Baldorioty's early activity consisted in writing scientific reports, using his analytic capacity to evaluate particular socioeconomic problems. He suggested the adoption of gold in absence of a common coin or '*moneda*', but studied many others as well. These include an analysis of guano in Mona island with Acosta, a study of gas counters or the conservation of forested hills. Baldorioty called for the creation of a university with true scientific training, and participated in a horticultural commission which identified the best fruits to be grown on the island.

---

[264] The pharmacy *Botica Babel* also served as a "*tertulia*", where problems of the day were discussed.

His most interesting report, however, was that of 1867 when he travels to the *Paris Exposition* shortly after Darwin's *Origin of Species* 1859. Baldorioty was particularly taken aback by the power of artificial selection, or the selective breeding over countless generations, and its power to improve agricultural stock. He immediately realizes its economic usefulness, as it would lead to meatier and larger chickens, a larger number of eggs produced, or an increase in milk production to perhaps 400 liters per year, doubling that of current production. His promotion of a zootechnical school was precisely based on such an experience. Expectedly, perhaps, the government foolishly rejected the proposal on legalistic grounds, taking up the entire front page of the *Gazeta*, the principal island newspaper, to publicly humiliate him. The school was rejected because Baldorioty lacked teaching certificates—in spite of his prior position in awarding such certificates in the first place.

Baldorioty was also attacked by local authorities for being a Yankee sympathizer. In his comments pertaining to the United States, he had recognized the importance of good laws. As with artificial selection, these had a powerfully positive impact in the development of a society when systematically applied over many generations. He recognized that the US financial success stemmed not entirely from its physical geography, but rather from its good laws, which sharply contrasted to the erratic nature of Spanish colonialism and colonial governors whose inconstancy of positions was socially chaotic.[265]

Jose Julian Acosta would visit his friend Baldorioty in *El Morro*, after the latter had been arrested during the *Compontes* of 1880. Being severely malnourished and in an increasing state of declining health, Acosta routinely complained to the authorities. But not much would happen. As the two were akin to brothers, Baldorioty's cruel treatment must have weight heavily on Acosta.

Upon their return from Spain, both had taught sciences, and played their part in numerous studies. Acosta's participation in a sugar cane commission was particularly important. Although the committee did not find the cause of the economically impactful

---

[265] No much has changed in Puerto Rico since then.

disease, it did suggest that the adoption of new cane varieties would ameliorate its effects; diversity was the best recipe against uncertainty. Both had also participated in the Spanish Court, and were equally denigrated for such participation. Acosta had been cofounder of the *Partido Liberal Reformista*, at different times serving as president. There can be no doubt that the two were political and scientific blood brothers.

However, their ideological paths were slightly different. In contrast to Baldorioty's moderate stance of autonomism, which itself contrasted to the armed revolution of Betances, Acosta took the more conservative route of assimilation. Although briefly arrested after the Lares revolt for 4 months, he believed Puerto Rico only required equal stating as a province in the Spanish state. Consequently, as president of the Liberal party, Acosta refused to make autonomist claims public—for which he is ousted from the party.

Acosta also attacked the pervasive limitations on liberty which existed in the island from a 'pragmatic' point of view. As Baldorioty's critique of the *libreta* system, his main argument against slavery was that it actually lowered agricultural production. Acosta used his position and intellectual standing to provide evidence which contradicted the government's official public stances and rhetoric. The key problem with many of the local laws and institutions is that they deprived men, both the owner and slave, of an incentive to maximize their work and output. Acosta calculates that slavery as a whole undermined sugar output by 33%. Worst still, slavery inhibited the formation of a market economy. He correctly pointed out that if slaves were to be given their freedom, through their wages they would become consumers of goods, and thereby stimulate the island's market economy.

It goes without saying that Acosta held a laissez faire stance a la Adam Smith, though he was not blind to its abuses. In contrast to Baldorioty, Acosta came from a well-to-do family whose fortunes drastically changed when Acosta was but a young boy.[266] As Baldorioty, Acosta showed a great deal of early natural talent and ability. At times he would substitute teachers at the *Seminario Conciliar*, and even requested to become a teacher but was found to be too young for the charge.

The case of Acosta is emblematic of the lack of professionalization of science in the island. As we have seen, positions in schools did not pay regularly, and teachers were often forced to supplement their income with extra outside work or '*chiripas*' as they are locally known. Acosta's father had 11 kids, and Acosta himself had 8 kids, and was thus quickly pressured into finding a source of remuneration. Being sharp minded, he imported a printer from Cuba and built a paper warehouse next door whose relationship was unclear to local tax authorities in the beginning.

Acosta eventually becomes one of the premier printers in the entire island, and played a substantial role in the Puerto Rican Enlightenment, as had the printing press in its European post medieval context.[267] Acosta's shop diffused all sorts of scientific works in island. One of these was a biological study of Andre Pierre Ledru of Puerto Rico, which Acosta stumbled upon while visiting a Paris bookstore. He also discovers Iñigo Abbad y Lasierra's monumental history of Puerto Rico, which had not been accessible to islanders at the time. His new journal, the *Almanaque Aguinaldo*, publishes many articles in what are now considered classic topics in the history of science, as a study of Johannes Kepler.[268] Other articles describe new technologies as the telephone and barometer, or biographies of historical scientific leaders from the Americas, as Franciso Jose de

---

[266] His father made a rather poor investment in a ruinous sugar plantation in Rio Piedras. To make up the losses, he decided to sell productive properties, which inevitably resulted in bankruptcy. He took his family to Ponce, and became a staid local notary.

[267] One of the particular features of the Renaissance was the exploration of new literature—in that case, the rediscovery of the Greeks via Arab translations. The low cost and superior quality of the printed works, led to the ample and rapid diffusion of knowledge and new ideas in the vernacular.

[268] Curiously, an article dedicated to Newton never appears, suggestive of the repressive context in which his business operated.

Caldas—one of the first true physicist of Americas. Another key feature of Renaissance humanism, the translation of classic works, can also be identified at the time.[269] The similarity between both periods is astounding.

As in Baldorioty's case, Acosta's moderate stance was seen as radical at the time, specifically his belief that the key to economic development was education. Only 20% of the population was literate. Acosta recognized that education allowed individuals to pursue their own self interest, and in the process provide new services and products for the benefit of society. As with Baldorioty, Acosta also sought to establish a university type institution in Puerto Rico, the *Instituto de Segunda Enseñanza*. After repeated attempts, the school is finally established in 1873 but closes soon thereafter in 1874. It is again reopened, with Acosta serving as its new director in 1882, but is again closed in 1884. Acosta repeatedly clashed with professors, most of whom were Spanish peninsulares. Upon being dismissed from the very institution he created, his students sought to get him back, but he refused to return. All that he was interested in, he tells them, was that they finish their studies. The dynamics of colonialism profoundly undermined the island's economic development and scientific progress.[270]

In conclusion, does science lead to sovereignty and nationalism? Did it foment the independence movements the United States or Puerto Rico? A clear and simple causality is hard to demonstrate, and the wide range in variation in the political stances of Puerto Rico's most important *proceres* is suggestive that a simpleminded relation does not exist.

There can be no doubt with regard to the scientific ability of our local characters. All had been trained in leading European

---

[269] Betances translated many Latin works into both French and Spanish. Some of these works included books by Petrarch and Plato. He was obviously trilingual.

[270] Thomas Glick argues that such conflicts, whereby educational venues of professional advancement were closed to creoles by *peninsulares*, spurred many creole scientist towards political independence. However, the does not seem to have had same effect in Puerto Rico, as Acosta was a lifelong 'conservative' and does not appear to have modified his ideological positions.

scientific centers. Acosta even traveled to Berlin where he met and interacted with Humboldt, who might have told Acosta about Caldas—an article of which appears in the *Almanaque Aguinaldo*.[271]

A rather diverse set of political positions in the United States also existed. Early on in his career, Franklin's political stance was more akin to that of Acosta's, and it is suggestive to note their commonalities. Both made a living from the printing press, and were very astute social men whom interacted with all members across their respective societies. However, had it not been for Franklin's negative experience and mistreatment by England, he might have forever remained a loyalist. We tend to erroneously imbue results of history onto the past as if they had been predetermined at the beginning of history.

We can point out, however, that the scientific ethos led to a much more critical and analytical evaluation of political circumstances. Political structures, particularly colonial ones, were not predetermined and but had to be constantly evaluated. In Puerto Rico, all enlightened *proceres* had been critical of the Spanish colonial regime in some form or other; that Acosta chose to hide his criticism in the work of Abbad y Lasierra does not negate its existence. It is certainly the case that they were not satisfied with the political status quo and Sanz was correct in suggesting that education in this scene had a noxious effect detrimental to the metropolis. As delegates of Puerto Rico to Cortes, their principal calls were for a change in the political status quo.

---

[271] Humboldt met Caldas during his travels in Latin America, and Caldas at one point even wanted to move to Germany. Humboldt himself saw Jefferson on his way back from his trip to the Americas. It has been suggested that Humboldt, as an Enlightenment man, played a key role in Latin American revolutions, but this claim that has been disregarded. He did interact with many in all social classes. However, Humboldt had to be very careful as he depended on state authority for permission to travel throughout the region. Acosta did admire the German university system, which gave professors such a great amount of freedom, the direct opposite to Sanz's position in Puerto Rico.

Perhaps one of the most important 'revolutionary roles' of Acosta and Baldorioty was simply that of scientifically undermining many of the state's irrational and arbitrary policies and positions, which had frankly become untenable. Their rigorous logic and accumulation of data demonstrated that many state claims were patently false. Their 'revolutionary pamphlets' were emitted via scientific reports, and perhaps did more to undermine the Spanish colonial system than Betances's continuous call for armed revolution—which for all sakes and purposes was ineffectual.[272]

Certainly, science in the United States was also used as an important political instrument. With it, Franklin helped the United States obtain independence, and once obtained, Jefferson used it to shape its institutions. Again, while science did not predetermine sovereignty, it played a big part in helping to define the democratic institutions which would insure it. The critical rational analysis of social institutions, led to their reformation.

In the end, science modified the intellectual context in which both monarchical and colonial institutions had been previously justified, and inevitably led to a revocation of the status quo. Science was ultimately liberating.

---

[272] Betances seems to have been more of a romantic in that his military plans routinely lacked rigorous logistics and were continually intercepted by government spies and snitches. Segundo Ruiz Belvis was killed in a hotel when he flees to Chile after the Lares revolt.

Euthanasia

Let us consider three hypothetical scenarios regarding the conscientious act of dying.

In the first hypothetical case, assume that you are gynecologist and a young pregnant woman has entered your office. She is nineteen years old and wants to have an abortion. Her boyfriend abandoned his responsibilities; some subcultures in the United States define manliness by the number of women they impregnate, without assuming any of the responsibilities entailed by such pregnancies.[273] The girl's family does not know about her condition, and so she had a very difficult decision to make on her own. She has actually considered entering medicine to become a gynecologist herself. Ultimately, she has decided to have an abortion, recognizing that the life ahead of her is a world full of wonderful possibilities.

Were she to have the child, her entire life will have to be dedicated to taking care of it. The possibility of higher education under such circumstances is difficult at best. She will require a great deal of discipline. She might perhaps succeed in obtaining a university diploma, but the odds are stacked against her. Higher education is hard enough, given the many obstacles in its course. With the baby, such an effort will be like swimming with a deadweight attached to her body. The outcome's default is failure, in that it will require 100% effort. While she might be able to succeed at first, any minor problem has the potential of wrecking her hoped for path.

She might have told you the story of her previous attempt in going to a different clinic, which pretended to be an abortion clinic but in fact was run by religious conservatives who propagandized that the ending of any life is a sin—no matter

---

[273] In transferring these responsibilities onto others, including the woman's' family, the state and public funds, this scenario is actually the exact opposite of manhood, discussed in a previous chapter.

what the costs or conditions of the pregnancy are.[274] When she met with a specialist, he not only denied her request for the procedure but actually made her feel guilty for the decision she was making, which was in and of itself already a very difficult decision to make. She had been fooled into trusting a professional who was not whom they claimed to be.

Should you carry out the abortion?

Assume now that you are now a geriatric specialist. A woman comes to you in a great deal of duress. She is a single child and a single mother as well. She and her husband divorced a few years ago, leaving the daughter with her. By experience, you know that it is not easy being a single mother, as the likelihood of entering poverty is very high. In spite of having a professional well paid job, the woman's life is still very stressful. With the recent death of her mother, she now also has to take care of her elderly father. The father developed Alzheimer's recently, and so the legal responsibility of tutorship has fallen on her shoulders.

Although her father is still relatively physically fit, as a person he died some time ago. He does not recognize anyone anymore. At first he had some recognition, but his cognitive condition degraded as the disease has progressed. He stopped remembering words, and his conversation has become incongruent; he can no longer follow the line of the conversation, and often veers off topic. Her father has lost his hold on reality. Although not a danger to others, he has started to physically decay as well.

The situation is increasingly becoming onerous for his daughter, who now cares for two 'infants', while at the same time trying to keep her career afloat. She has to clean and bathe her father like a baby, which is emotionally difficult as a female daughter. That her father no longer recognizes the daughter is perhaps the most traumatic aspect of his new condition.

When young, the two had a very special bond and relationship. The mother had become an alcoholic and was often abusive. Under the circumstances, the father had taken the difficult option of divorce and won custody over his only

---

[274] They do not consider if a woman has been raped or deeply abused.

daughter. The father had become the de facto single parent of the house, and, as the only female in the house, the situation had become challenging for both. She remembered how he had been forced to deal with sensitive female issues when she was a teenager; although they made him uncomfortable, he had successfully managed them. In short, as a physician you become aware that there is a long and tender bond between the two, which made her decision that much more difficult.

Upon arriving, she requests that her father be euthanised, as she did not have the means to take care him, or to place him in a home. In the meanwhile, she has reached the limits of her endurance. She was tired and worn, and simply could go on no more.

What do you do?

Finally, in our third case, consider that you are a veterinarian who runs a successful clinic. A man walks in with a sick and elderly dog. You notice that its abdominal area is distended, likely from cancer, and that the dog is in a great deal of pain. The dog now yelps and bites when others approach him. When the disease first started, the dog was relatively functional and could walk around. As the cancer progressed, the dog lost its back leg functionality, and was now leaving excrement and urine all over the house. The man wants you to euthanise his dog.

The dog however seems to be relatively happy. For years the dog had played an important role in the family. As a biologist, you are aware that the bond with canines is due to a dual benefit, holding a symbiotic relationship. The house was in a relatively dangerous high crime area, so the dog acted as the best alarm system there could be, barking at anything that moved and thus quickly raising an alarm from any potential threat. You are aware that if a thief has to pick between two houses, he would rather pick the house next door because the relative cost of entry was too high. Statistics indicate that the sheer presence of any dog substantially increased the security of its residents.

In spite of all of this, the man has asked you to euthanise his dog, as the Humane Society was no longer euthanising dogs in spite of the abundant animal problem in the area. The owner considered simply taking a machete or gun and killing the dog, but believed that his faithful dog's life required a dignified ending, if not for all the years of service dog had provided for the entire family. He knew that the most dignified manner of death was via a veterinarian, whom with a simple injection could put the dog to sleep in a manner that was painless and merciful to the animal—as if going for a very long sleep. As a veterinarian, you were well aware of the fact that if you did not grant euthanasia, the owner might be forced to violently end the faithful dog's life; which would be traumatic to both dog and owner.

What do you do?

All three cases have various common features. All three cases deal with the euthanasia of ideal exemplars of their kind: the faithful dog, the good father, and the innocent embryo which has done no harm to anyone. These cases at first might suggest that the moral integrity of the potential victim should play a role in the final decision making, which raises the broader question whether one should assume that life is to be saved under all and any circumstances. As a physician with the power to render death, should you comply with your patient's requests, either to themselves or onto others?

There is a principle of the law relevant to these issues, which unfortunately is routinely ignored in many legislative decisions: the principle that authority should always go hand in hand with responsibility.

If an individual had been given a particular responsibility, it consequently then follows that they should have the necessary powers to comply with their respective tasks. This notion should be a guiding principle of all law as it is both fair and just—but unfortunately is seldom observed. Its breach, curiously, touches on every person s 'unjust senses', as its failure of implementation traps individuals into responsibilities which can never be met as they lack the means to comply with these. It is no different perhaps from slavery, and is a bad deal for any individual to be involved under such condition given its high probability of failure. Lacking the authority necessary to comply with one's responsibilities also means that one's reputation will ultimately be harmed, given the inevitable failure. Hundreds of such cases occur every day in Puerto Rico.

Rodrigo Fernós

As discussed in prior cases of rape and cuckoldry, these situations amount to the elimination of human agency. Think of the veterinarian; as they do not hold the responsibility of caring for the creature, the decision of euthanasia is not theirs to make. Yet laws do not exist in Puerto Rico which force a veterinarian to undertake the euthanasia upon the owner's request, which simple consideration reveals what a failed social policy this represents. By raising the cost of ownership in this manner, the likelihood of adoption within the context of an oversupply of domestic animals roaming the streets is reduced.

The same thing can be said with regard to the requested teenage abortion. Ultimately it is the woman's decision whether to carry the child or not, for the ultimate responsibility for the rearing of the child will fall on her shoulders. The burden of being a caretaker is not a light one, and could also have grave social consequences, in this particular case clandestine abortions which can result up in death of mother. The problem of overpopulation does not exclusively reside within domestic animal populations.

The same might be said in the case of the elderly father. If the cost of the imposed caretaking is too onerous, the caretaker might be forced to resort to actions which will traumatize all individuals involved: the violent taking of life, and the psychological trauma it implies. As with much else, life is not of an absolute value but rather a contextual one that varies. Worst still, the highest medical costs in the United States are incurred at the end of life; the exponential rise as a percentage of GDP points in the direction of financial abuse of countless senior citizens. The insistence of keeping a loved one alive, regardless of all circumstances, follows the law of diminishing returns[275]: increasingly high marginal costs with few benefits.

Many in modern Western societies often find it hard to deal with death, in spite of the fact that at some point, we will all die. When we are younger, we tend to implicitly presume an infinite future ahead of us—a future characterized by the hope of uncertainty. In the past, however, death tended to be a common

---

[275] The term in Spanish is *ley de rendimientos menguantes*.

experience as medicine was not as well developed. However, as medicine has advanced in the bacteriological revolution of the nineteenth century, it created a 'demographic transition' and to changed expectations: less babies and longer lives. The emergence of modern medicine created new social forms and new types of social relations, in which death came to be perceived as an aberration. There is an implicit modern presumption that we will live forever, so many are willing to take extreme measures to deny that which is humanity's fundamental truth: our own mortality.

Computers in Biology

The reader may have noticed that the role of information in ethics has been a continuing theme of the book. We first looked at various biological theories of ethics, as social Darwinism with its emphasis on competition or eugenics which saw reproduction as the key to development. Sociobiology, the most sophisticated body of ideas discussed so far, in turn gives information a special role. Our evolutionary background sets us in a social context, which itself influenced the evolution of the human brain. Specifically, our concern with reputation and cheating overwhelms many other traits, as aptly noted by Robert Axelrod's 'the long shadow of the future'. Information plays a significant role in the moral conduct of man.

Both Japan's Unit 731 and the Guatemalan Syphilis study, for example, were gross violations of moral and ethical conduct by any measure. Consequently, there was the enormous amount of effort placed to control and conceal information that emerged about these to the general public, leading to a strange almost-fictional conspiratorial dynamic. All who participated sought to prevent information of its existence from leaking to public. Somewhat shockingly, they were almost successful, had it not been for the accidental discovery by a historian (Guatemala) or the willingness of culprits to testify almost half century after events (Unit 731).

It is clear that public information about misdeeds is perhaps the greatest threat to immoral behavior—to the point that these enter a vicious cycle resulting in even worse criminal actions. That so many young women often go missing suggests that these were the victims of rape, and represented potential liabilities for their perpetrators, doubly suffering as a result. Even the most bloodthirsty of criminals cherishes their reputation, for holding a good reputation has enormous social benefits—the loss of which implies large repercussions and substantial losses.

That being said, the moral outcome of information is rather complex, and it should not be presumed that its presence will automatically result in proper moral behavior. The relationship between information and ethical conduct is not a simple and direct cause-effect relation, a good example being the Baltimore case. A Noble prize winner was accused of fraud. Though it was not of his own doing but rather that of a colleague's, David Baltimore was willing to engage in a decades long fight to justify the unjustifiable. He appears to have been so fearful of a degradation of his academic standing, that he was ultimately willing to use the enormous social stature that comes with being a Nobel Prize winner to defeat his accuser: a humble and lowly postdoc Margot O'Toole. His actions are as reprehensible as the rapist who murders his victim. Baltimore would have continued fighting ad infinitum had not scientific journals finally gotten fed up and placed a stop to the publication of articles relating to the story.

The reader may also have noticed that a small amount of space has been dedicated to a formal definition of what is ethically correct. Sociobiological theory indicates that we tend to be moral in the vast majority of cases. We know what is correct because we have an evolutionary driven innate sense of justice, with the exception of psychopaths who can be defined as individuals with damaged brains, and hence outside our purview. The vast majority of people know right from wrong, 'in their heart'. Whether they act on such feelings, however, tends to be affected by external pressures, which is itself another issue altogether.

Benjamin Franklin used to say that all liars are easily detectable, and he is correct. The brain not made to hold multiple variants of a story at the same time, which inevitably leads to inconstancies in the liar's story telling. The brain is not made to retain a thousand variations, as well as to identify which individual received which variant. The detection of the irregularity by a single recipient is simple, as they only have to recall the prior story to identify the inconsistency. The cohesiveness of each variant also becomes distorted, making it impossible for the liar in the long run to succeed in lying convincingly.

Given the significant role of information in ethics and human behavior, it is perhaps useful in this final chapter to study the role of computers in biology. It implies that computers also have significant role in the ethical behavior within a community—particularly so in our 'postmodern society where computing power and digital information is so pervasive. Technology has enormous transformative social effects, and neither biology nor ethics are exception to the rule.

Computers and its derivatives as smartphones have become the sine-qua-non of our postmodern age. Their ability to quickly store, retrieve and calculate make them much more than bicycles of the brain as Steve Jobs used to say. The metaphor presumes an image of greater power and speed without altering the essential character of human cognition. Yet computers alter our relationship to information.

As with so many human dynamics, these relationships are torn between two fundamentally different visions, one which is a top-down 'dictatorial' model with little user agency and the other tends to be 'democratic' in the promotion of information which flows from the bottom up, and is abundant in user agency. Our acceptance of either model is likely determined by our personal cognitive mode. Conservatives as a rule dislike ambiguity and chaos, which liberals are more comfortable living with uncertainty. Simple solutions often tend to disastrous consequences.

As both technologies for the creation and manipulation of information, computers have also become our principal means of communication, and thus it is useful to turn our attention to the impact of prior technologies—in particular the rise of the postal office, which served the same function and is increasingly becoming the symbol of a bygone era.

With emergence of computers and telecommunications in the twenty-first century, postal office use is in steep decline. Through much of the twentieth century, long distance contact could only be established either via phone or letter writing. However, long distance calls used to be incredibly expensive, in today's terms around $50 for a 10 minute call during the 1970s, making it cheaper and more effective to send a letter. Today, much of this is now done via either email or texting-video chat, which have such low rates they are almost free on a per message basis. This decline in use has removed a key source of revenue from the *United States Postal Service* (USPS) and has forced a restructuring of the government agency.[276] That being said, it is important to briefly look at the post office's emergence and the social consequences it had, to get a hint of its impact and the ethical role of computers in postmodern life.

Prior to 1845, the postal office existed but the amount of letter writing was minimal. Its main function was that of distributing newspapers. As information was deemed essential for democracy by the US founding fathers, the 1792 Post Office Law was established which subsidized the delivery of newspapers and other print media.[277] As late as 1853, 7/8 of the USPS total delivery weight consisted of newspapers, even though it provided only 11% of its income. The USPS has traditionally been run with operational deficiencies. Inversely, the cost of mailing a letter was initially prohibitively expensive, consisting 30% of a single day's salary. This high cost meant that only businessmen used the postal service, as they were the only ones who could afford it. Banknotes, seeds, and daguerreotypes were commonly sent in such letters.

One benefit of the system, however, was that that all US citizens received newspapers. By contrast to today's uneven distribution of telecommunication services, all rural areas received news of some sort or other.[278] A consequence of this

---

[276] In 2015 alone, the USPS lost $5.5B, and between 2007 to 2010 $20B. Since 1971, the operational losses amount to $47B, of which stem from the drastic decline in mail volume. In five years alone, male volume fell from 210B pieces to 150B, or nearly a 30% decline, which would be devastating to any business.

[277] Both post offices as printers were served well by the bill, including Benjamin Franklin.

[278] We can see the emerging tendencies by Bell Company policies, which tended to refuse provision of services to rural regions for being too costly given the high dollar per mile investments. Such arguments are spurious when seen in historical context. Bell's policies were detriment of farmers, who in

system was that initially letter writing was often done via newspapers, as a way to keep in touch and let others know you were well. The sender might buy a newspaper, and underline certain characters to send a message; it was a creative response with clear limitations.

However, in 1845 the US postal services underwent a reform, which led to a drastic increase in letter writing. Economic studies revealed that high volume transactions would be able to cover the costs of the system. Lowering the price of stamps meant a greater rate of use, and in turn an exponential increase in volume which would pay for itself. Postal rates were lowered to 25% of their original price, and as predicted were followed by a drastic expansion—so much so that postage stamps often were turned into currency by its users.

Letter writing exploded in urban areas. By 1854, the sixth largest cities had mailing rates that were six times those of the rest of the nation. Manhattan alone produced 10% of the entire continental postal system. This change also led to a revolutionary modification of expectations. Nathaniel Hawthorne commented on the sheer miracle that an anonymous letter would arrive at its intended destination, a function which today we take for granted in our golden era of electronic communications.

The postal system's social impact was most noticeable during the Civil War and the California Gold Rush, both of which produced an enormous volume of letters. Every single day during the Civil War, some180,000 letters were exchanged. The lines to the post office were be so large, that senders were willing to pay $50 just to cut in line—an amount equivalent to $776 today. So significant was the receiving and sending of letters from loved ones, that a transformation in personality occurred. Men who were ready to kill each other at a moment's notice, would suddenly become the best of friends after receiving their mail. The trauma of the war was made patently visible, given the countless stories of the most violent of men who would break down like little children upon receiving letters from home. Letter

---

turn tended to set up their parallel telecommunication systems using cattle fence lines for their transmission.

writing had become a way of extending the traditional nuclear family, which at the time was also undergoing a profound change.[279]

Letter writing thus allowed for the maintenance of family contact and relations, which had a moralizing function, which is perhaps well described by Chilean philosopher Valentin Letelier; daughters, wives, and mothers, women reminded men of their humanity. Letter writing also led to the establishment of the norm of privacy in communications. One clause of the 1792 Postal Office Act was that postmen could not open letters, a requirement that probably originated from business interests. As letter writing became common and pattern of daily life, the expectation of privacy spread through the nation; the government could not and would not intrude upon the private affairs of its citizens—a distinctly protestant notion.

Print was specifically regulated by the first amendment of the US Constitution, and it is interesting to notice the particular value placed on the freedom of expression by US Founding Fathers. 'Thou shall make no law' barring freedom of speech, served as an essential requisite for a democracy. All should be allowed to express themselves. Were the founding fathers sociobiologists? While obviously they were not, as the science had yet to be discovered, they were very perceptive of human nature, Privacy, however, was circumscribed under certain restrictions—in particular the clause of "clear and present danger" as established by Justice Oliver Wendell Homes. One could not shout "FIRE" in a movie theatre as it would lead to a stampede, panic, and the likely loss of life. While the first amendment is important, it was not absolute, and herein we see the thoughtful and careful balance of social needs and forces.

The threat of prior restraint could also not be used to violate freedom of speech, for it was the tendency of governments to control information within their borders, which in turn helped insure their powers. This clause was meant to prevent vague wording in laws which might contravene the first amendment—a

---

[279] As the US was modernizing, maturing children were forced to leave home to look for work and employment elsewhere, which was a relatively new experience in US history.

typical abuse of local state power. The US Founding Fathers preempted mechanisms which might have been put into place by states wishing to overrule constitutional restrictions. Unfortunately, they did not anticipate the rise of the corporation in twentieth century America.

As the United States industrialized and science began leading to new applications, one of the problems which emerged was the concurrent but distinct regulatory legislation of new technologies which ran parallel to traditional modes of education as letter writing. The printing press as a technology had been greatly valued by the founding fathers, for which institutional protections were embedded. However, with the emergence of the electronic realm, new regulations emerged out of their own historical dynamics and independent of the rules which had governed traditional written communications. The telephone law, for example, was based on prior telegraph law, which in turn had been based on prior principles of railroad regulation. Here, the motivations were primarily financial, as monopoly power could destroy markets, rather than philosophical considerations as freedom of expression.

In the specific case of railroads, the principal concern had been the creation of a vertical corporate structure, as demonstrated by Rockefeller Standard Oil, whose control of railroads—the main distribution lines of petroleum at the time— could prevent the delivery of oil by their opponents, thus driving their competitors into bankruptcy, leading to abusive monopolies. As a result, early telecommunications law was based on the common carrier law. As phone companies were monopolists by nature, they were hence subject to regulation by state, and were specifically prohibited from altering the rate and flow of information. Early telecommunications law was thus established within the mentality of a market context rather than that of constitutional principles. That these principles might be achieved was but a secondary and accidental benefit of their commercial regulation. There was no clause, in light of the absence of explicit limitations, which prevented the violation of the US Constitutional order, as noted by Ithiel de Sola Pool.

Telecommunications regulatory principles were thus early in their history based on economic dynamics rather than on philosophical considerations, and would be consequently applied to all other electronic communications technologies that followed thereafter—as the radio. In the process, diverse codes of regulation emerged which varied according to the particular technology involved, but which shared no necessary consistent legal principle. This did not represent a problem initially given that the early history of the industry was characterized by many small companies and radio enthusiasts. While nobody would ever dream of owning radio waves—the notion sounded like the mad idea of trying to own the air—views began to rapidly shift as telecommunications advanced.

The potential of immediate mass markets soon emerged, with the radio's ability of reaching thousands of people instantly, which in turn implied a great deal of social power and advertising revenue in the creation of the mass media. There was a consequent corporate push for regulation of the industry, formalized by Herbert Hoover during the 1920s. Under the new regulatory order, distribution of the signal was to be based on the principle of scarcity. Since there were a limited amount of bands and frequencies available, all radio stations had to give air time to alternative views, and in principle helped to preserve constitutional safeguards. It is interesting to observe how institutionally proactive the US tradition historically has been.[280]

Corporations played a substantive role in the rise of the electronic age; their growth was of such a degree that they were forced to develop key precursors to our information age.[281] Prior

---

[280] A conflict with prior principles is identified, thus a law is established to explicitly resolve these. Notice that these institutional solutions were not just subject to market forces and corporate competition, but rather by clear legal boundaries that were fair to all views—in spite of fact that did not know what these views actually were prior to their expression.

[281] The modern corporation is result of legal changes towards the end of the nineteenth century. During most of US history, they tended to be small affairs, and were never mentioned in the US Constitution. The US Founding Father were certainly not aware of the beasts corporations would become. Originally, the traditional joint stock company was formed for very risky ventures with definite social benefits. In essence, stock represented non-guaranteed loads

informal forms of management, such as the issuing oral haphazard instructions that were easily forgotten, became chaotic by the end of the nineteenth century. Managerial information stored in bounded volumes could not be easily handled, or quickly adapted to new demands. Corporate functional requirements gave way to systems management developed by engineers. Apparently minor changes drastically improved efficiency, as the use of 3x5 card filing system which could be quickly rearranged and searched. The use of typewriters meant that easily and consistently readable information could be generated quickly, at 120 words per minute rather than the former 25 words per minute. Vertical filing systems increased the amount and density of information stored. The number of workers in this new clerical employment as stenographers and typists quickly grew from 33,000 employees (1890) to 786,000 (1920). New employment opportunities emerged for females; by 1900, 74% of clerical positions were held by women.

Corporate demands at the turn of the century thus formed the origins of the information society in a context of a paper based control-command structure—which would also come to characterize the early world of computers.

The rise of electronic technologies as the telephone created cultural challenges for communities as the Amish, wishing to retain their traditions in spite of all the technological changes surrounding them.[282] Amish identity is specifically anti-modern, rejecting modern secularity and all it entails. This rejection led to split with the Mennonites, who were more willing to accept the amenities of the modern world. Although their rejection of modern clothing and pockets led to the Amish distinctive cultural look, deep internal divisions also existed within the community. The Amish opposed telephones as it represented an external intrusion from the secular world.

---

issued by citizens to an organization. If the venture failed, the stock holder lost his investment. In this, corporations in their early forms were rather courageous enterprises, contrasted to today's institutions that insist on guarantees on profitability.

[282] Amish towns are typically composed of small communities, some 6,000 individuals throughout Iowa, Ohio, and Pennsylvania.

The diffusion of the telephone thus became a difficult cultural transition at the beginning of the twentieth century. There were obvious benefits to the phone. The increasing complexities of life in the US meant that Amish families were also growing and becoming increasingly fractured. The consequent increases in land prices forced many youths to move to urban areas. The Bell Company also tended to favor urban areas over rural ones, as the cost/benefit ratio was much higher in rural areas. Their absence stimulated the formation of many local private companies, which Bell oddly attacked—even in Amish communities. [283]

As the phone represented an intrusion from the external and foreign cultural world, they were not originally allowed within homes. Telephones tended to be placed as 'outhouses', in booths apart from homes. However, when some individuals did obtain a phone, it led to a particular moral question. If it was immoral for an Amish member to own a phone, was it not also immoral when their neighbors borrowed their telephone lines? The more liberal Amish were puzzled at the hypocritical contradiction of being attacked. Ironically, those who transgressed values and social norms were subject to 'shunning' or social ostracism, and cut off from local social information.[284] The phone represented a genuine conundrum for many families.

However, the problem was identified within its respective value context and a conscious solution was established for the broader collective. It was finally resolved that all could own a personal telephone, with the exception of religious leaders as these were typically selected for their exemplar moral leadership. This solution was seen as a form of protection of ideal models while retaining personal technological benefits, explicitly reconciling technology with the value system of a community. Technology was not something that irrevocably altered the values

---

[283] Such practices led to their reputation as a type of octopus, with its had hands in everything. In one instance, bell sent representatives to an Amish gathering prevent the formation of anew company but because everyone knew each other they were easily detected. Bell's efforts backfired, and the local company was established by local Amish.

[284] There are various levels of 'shunning', the worst forms is when done by all members of group.

of society, but rather the result of a conscious and controlled adoption. The same could be said for the evolution of computers.

We can divide history of computers into two broad periods, the first between 1950 and 1970, dominated by mainframes, and the other between 1980 and 2000 with the dominance of the personal computer. These are sidelined by transition periods, as the rise of the cell phone towards the new millennium. The two periods had distinctive cultures.

Mainframes were top-down machines. These enormous machines were usually leased to corporations, which could only be serviced by the seller's own technical personnel. It was legally forbidden to open up a mainframe for self-repair—even if such a task was possible. This legal landscape created a great deal of pent up demand, particularly by technicians who fixed computers, as all wanted to own a computer. Various factors came into play at time.

The *Home-Brew Club* created an open ended forum for computer enthusiasts, resulting in the first ever 'personal computer' in 1975. The *Altair 8800* was very primitive by today's standards; it had no monitor screen, and one had to switch levers to input data. The *Whole Earth Catalog* formed by Stuart Brand also had a substantial influence; though it did not make much money, it promoted the values of hippie counterculture. These were the complete opposite to those of the 1950s, well described in Herbert Marcuse' *One Dimensional Man:* the company man who blindly follows orders. Marshall McLuhan and Buckminster Fuller were also influential at the time, endowing the period with optimism.

Their principal critique of mainframe culture was that it subverted humanism. Human beings were not the means of technology but rather its ends; as a result, the key question that had to constantly be asked was whether technology could be changed to serve indeterminate human needs and purposes. Technology did not dictate the goals, but was rather only a means to an end. The common fear of nuclear war and Soviet take over of the US also generated a great deal of anxiety at the period, as it represented a substantial shift in the American way of life. Mainframes in this context seemed just as menacing, and impacted men as Steve Jobs, who regularly read the *Whole Earth Catalog*. He decided to take a consumer-centric approach, even if this vision today has become grossly blurred. Jobs and Wozniak create the Apple II computer, the first mass market computer, and both became new Silicon Valley multimillionaires.[285]

The Cold War also set the context for the emergence of the internet. If a nuclear war were to begin, would the US telecommunications network survive? The answer was resoundingly negative. If certain nodes were hit, the entire US telephone system would collapse—particularly impacting the long distance carrier ATT. With the establishment of TCP/IP packet exchange, any node could be destroyed, but the entire system as a whole would continue to operate.[286]

As most of us personally experienced, the internet had a marked impact during the 1990s, begun perhaps by the creation of http protocol by Tim Berners Lee in 1990 and the creation of *Mosaic* browser, later *Netscape*, which led to an explosive growth of internet use by 1994. The mindless information chaos that followed soon thereafter was resolved by Google, whose search engine appeared to be miraculous relative to its predecessors, as Altavista.[287]

The majority of computer in 1950s had distinct features. As a rule, mainframes were very costly and as a result compute time was typical shared between different terminals. The user had to

---

[285] The *Apple II* computer had a program no other computer had: *VisiCalc*, which was an enormous hit (today called a spreadsheet). *VisiCalc* was 'killer app': an application so important that consumers would buy machine just to run the software. Accountants almost fainted when they first used it; what took weeks or months now could be done in minutes: calculate how prices in one affect others, thus help calculate ideal points. Ultimately this would be rather ironic, as Apple now produces close ended systems, which mimic the culture of mainframes. Apples does not like owners opening computers and making any sort of upgrade, enhancements or improvements that Steve Jobs had been so critical of IBM, particularly in the1984 Macintosh commercial directed by Ridley Scott.

[286] Both Vint Cerf and Paul Baran invented the packet exchange, independent of each other.

[287] Larry Page and Sergey Brin actually tried to sell their innovation to various companies. When these did not accept, they begin their own company (Google), and thereby revolutionized the world.

wait for the computer to cycle to its answer, which would then be printed back. Each mainframe had their own operating system, unique to the purposes for which it was used.

One of its earliest uses was for airline reservations, the SAGE system, which was very complex but perfectly suited for the task. It was able to coordinate flights all over the nation, get inputs from multiple places, and provide a coherent and consistent output at a magnitude and scale that is still today impressive. The IBM 360 was another important innovation in that it created a unified operating system over a range of small and large machines that were mutually compatible. This allowed for a much greater scalability not found in prior systems. The IBM 360 was readily adopted in large nationwide businesses, and again conformed to the prevailing ideal of corporate command and control.

In spite of its capacities, the prevailing mainframe computing systems of the 1950s had egregious flaws—worsened as the computer increasingly became an integral part of the US economy. The company who acquired such mainframes could not repair them, and would have to wait for IBM technicians to arrive and fix the system. While obviously it was in the interest of IBM to quickly repair any issue, the arrangement introduced unwanted dependencies, particularly so with its most important feature: business information.

It is curious to point out that in 1940s, ATT Bell Labs had introduced the transistor, which as we all know formed the backbone of the computing revolution, but was mainly used to reduce their dependency on telephone operators—some of which could be 'angels' (assistive) while others were seen as 'demons' (obstacles). ATT had become the largest global corporation with 1 million employees. In 1968 microprocessors introduced, whereby thousands of transistors were placed on a single plate. One of the first integrated circuits was the 8008 , followed in 1974 by the Intel made 8080: 8 bit, and lower power. These integrated microprocessors formed the basis of the first personal computers.

It is somewhat paradoxical to consider that the company that would usher the personal computing revolution, was a classical vertically integrated company that leased telephones to its clients. Worst still the 1956 Consent Decree by the Justice Department, prohibited it from entering the computing industry—even if it could produce internal innovations. In 1965, ATT with GE, MIT, and others launched the *Multics* initiative, a huge mainframe service with 1,000 terminals, allowing 300 users to work at the same time. By 1969, ATT withdrew, as it had become too costly a project. Kevin Thompson by then was working at Bell labs, a brilliant computer scientist who studied at UCAL (Berkeley).[288]

Thomson disliked the IBM System 360, as it had a poor design for programming and too much bloated software.[289] Thompson brilliantly sought to recreate *Multics* in a single computer after failure of the project in 1969, hence the original name of *Unics*, which was later changed to *Unix*.[290] As Thompson and his colleague Dennis Ritchie was not pressured by commercial demands, his final product came out 'perfect', and, best of all, did not require enormous computational power to run.

Unix was thus used to run small PDP7 minicomputers, with 4 kilobytes of memory. Though they cost $72,000, they were still relatively cheap for the time, as well as the PDP11 of 1971—the most popular minicomputer with its rich input/output (I/O) tapestry represented expandable functionality. Some 600,000 models were sold, and revolutionized the prevailing assumptions dominating the world of computers. There were many other important innovations at the time. While in 1975 the personal

---

[288] Thompson actually wrote the first *Asteroids* video game.

[289] At the time companies focused on the number of lines of code. The longer a program, the better it was believed to be. Estimates of value were based on a count of per thousand lines of code. One would speak of 20k or 4.5K projects, for example.

[290] Thompson was brilliant. He wrote the operating system in one month while his wife was away on a work trip; week for kernel, one week for shell, one week for editor, one week for the translator/assembler aspect. The brilliance of the UNIX system was that it was mad up of little text programs, which could each stand separate from one other, but feed information from one to the other if need be. This structure was immensely powerful and adaptable, and a brilliant demonstration of elegant programming. Thompson in fact, did not have terminals at time, but would rather write his programming down on paper.

computers did not exist, by 1980 it represented a billion dollar industry. The information sector of the economy grew drastically relative to others by 1980: agriculture declined from 90% (1900) to 2%, manufacturing fell from 40% (WWII) to 22%, while the information sector grew to 46% by 1980.[291] It goes without saying that the information sector continues to grow to this day.

One of the sciences most affected by computers was biology, in which genetics is still undergoing a quiet revolution. Genetic data only makes sense in the light of computation. Its key elements, A-T-G-C, represent an endless series of amino acids which are run through complex algorithms; without computing, genetic information is worthless—hence the prominent role of computers in biology today. So predominant has computing become in biology, that some suggest it might lead to a possible redefinition of biology itself.

These changes can be readily observed in the overall structuring of its facilities. In the traditional laboratories of the 1970s, a key distinction was made between 'office' and 'lab' space. The laboratory symbolized what today most would imagine to be a stereotypical lab, full of test tubes, pipettes, and so forth. The lab was also characterized by a great deal of collaboration amongst its respective scientists. However, today, laboratories are distinguished between 'wet' and 'dry' types. While the wet lab is the traditional lab, the dry lab is made up of rows upon rows of computers and servers. In the latter, individuals do not interact with each other, as the main source of interaction is with the screen.

Walter Gilbert of Harvard, whom we have previously encountered in the O'Toole-Baltimore Case, is the creator of gene sequencing and has become profoundly worried about the impact of computing in biology. One of his concerns is that biologists will be transformed from scientists into technicians, no longer asking profound questions except those pertinent to grant

---

[291] The change even had an impact in economic theory, in that previously only tangible goods qualified as having economic value. The rise of the services industry also intruded into economic thought, and came to be defined as an important part of the economy. Today it is the largest growing segment by all measures.

awards. The concern is well illustrated in the history of the Broad Institute.

The Broad Institute in Boston (MA) has a dual facade. The 77C building is an impressive seven story building covered throughout with glass where young doctorates can routinely interact with one another; there is a 'Google' type environment at 77C. This arrangement, in which biology PhDs are the most valued employees, was used for publicity purposes, and resulted in a $400M grant. All internal windows are transparent, as if to suggest 'we have nothing to hide', in spite of the revolutionary implications of the genetic work there conducted.

However, another feature of the Broad Institute is hidden away at 320 Charles Street, where a very different dynamic occurs. It is essentially an industrial center, with 120 machines worth hundreds of thousands of dollars each, continuously churning out digital DNA. The machines process the DNA even if they do not automatically decipher the code; identifiers are used to indicate varying levels of uncertainty.[292] It is so costly to obtain information, that any information obtained is deemed to be valuable—in spite of its large amount of ambiguity associated with its byte.

Biology at 320 Charles has become 'industrialized' in various ways. For example the floor platform was studied using Taylor analysis so as to minimize the amount of physical labor. Psychological profiles of the workers made, and these were incentivized to work at the facility. For example, they were given internal promotions to raise level of loyalty. The hidden building is also characterized by an ethnic split in its labor force. While all workers at 77C are mainly white Anglo-Saxon PhDs, the workers at the 320 Charles are typically minorities just out of biology undergraduate programs who do not realize their personal work will never directly result in revolutionary discoveries.

There is a hidden battle with regard to the value of computing and genetics in biology and the term itself *bioinformatics* is under current dispute. With the rise of the first digital computers, the NIH initially viewed bioinformatics as merely digital archiving,

---

[292] For example, R= G or A; B= G, T or A; and N=all of the above (ATGC).

and did not appreciate its full potential. This disparity between policy and potentiality can be tragically seen in the biographies of early participants, as the cases of James Ostell and Margaret Oakley Dayhoff attest.

Ostell had actually developed programs for genetic sequencing during his doctoral work, but as they were not viewed as a true scientific contribution, he was forced to follow his wife to Vermont where she became a physician and principal income earner of the family. Yet the programs he wrote in Fortran were immensely popular, and he was continually deluged with requests for copies; at some point he ported his application to the Apple Macintosh platform as the *MacVector* app. His user-friendly programs were of immense benefit, quickly giving users the information they needed, while other programs began with a typical typically cryptic ">?" in the first line.These applications helped countless other researchers in biology, but were still unfarily not formally considered a direct contribution to biology.

Just as Thompson at Bell Labs, Ostell began designing programs for minicomputers, whose cost by the late 1970s had fallen to the $10,000 range, and held 46 kb memory and 8 bit Intel microprocessors. At the same time, the number of scientific articles with the term bioinformatics began to drastically increase. First used in 1975, its frequency increased to a hundred citations per year during the next decade, exploding after 1992, to more than 10,000 papers published per year. Although there were some 21 university programs, by 2004 these grew to a total of 74 bioinformatics programs.[293]

Eventually Ostell applied his program to a particular case, for which he was finally awarded with a formal doctoral degree. The degree itself showed that the general perception of the area within biology had changed. During the 1970s, the notion of a biologist knowing computer science was very rare; yet by the 1990s, it was presumed that all needed to have a fair degree of computing under their belt. Again, a gene can only be understood in a

---

[293] Some claim that 'bioinformatics' will disappear as it is gradually incorporated into biology proper, under the belief that it merely represents one more instance of technological euphoria in the United Sates.

computational context, so the increasing rise of genetics was by definition accompanied with a rise in the use of computers in biology.

Margaret Oakley Dayhoff's story is much more tragic. Dying in 1983, she was a pioneer in the use of computer in biology, placing the understanding of life's evolutionary tree on firm grounding.

Her doctoral studies had pushed her into use of computer in the very early days of computing in post WWII US, where she used tabulated cards to solve computationally intensive problems. In her collaboration with Linus Pauling, she realized that protein structures could provide evolutionary maps, so she created a matrix of point mutations (PAM) to identify such changes. Not all proteins will show the same rate of change, each gene having a different probabilities over its history. Her PAM matrices established the likelihood of change, which could be multiplied by themselves to establish change over countless generations, ad infinitum.

The work culminated in the *Atlas of Protein Sequences and Structure* (1965), which provided an unquestionable and unique view of evolutionary history. Regrettably, many discredited her work because its original data had been taken from the research of other biologists. At the time, the definition of a biologist was to produce the raw data oneself, so in spite of her contribution, she was denied the nomination for entry into the *American Society of Biological Chemists*. While it is certainly true that the raw material of work had not been gathered by her, it does not undermine her contribution, as it occurs on a higher level order of abstraction—akin to the work of Georges Cuvier.[294]

---

[294] Cuvier's revolutionary work established a classificatory structure for fauna, akin to that which Linnaeus had established for flora. It was based on the comparative anatomy of countless specimens gathered by others. Cuvier was forced to respond to the same critique which would be issued two centuries later towards Dayhoff. Cuvier noted that, the natural historian would not subject to fleeting impressions in the library, but could systematically and calmly reach broader levels of generalization that were impossible for the field naturalist.

As the entry of computing into the world of biology intensified, a number of groups began issuing unsolicited applications to the NIH in 1982 for the establishment of a genetics database, which would eventually become *GenBank*. That such a large number of applicants were seeking funding for the same project suggest the clear convergence of various factors, including the continued development of genetics and the exponential increase in computing power and storage.

Unfortunately, Dayhoff did not receive the $3.2M grant (over 5 years), but was rather awarded to Walter Goad at Los Alamos. Goad was a physicists whose doctoral research had been based on the study of cosmic rays—an analysis that required statistical mathematics of the sorts similar to that used in the creation of the hydrogen bomb, of which he participated.

As one cannot trace individual particles, Goad borrowed formulas from fluid mechanics, also extending these techniques to the field of medicine. Again, as one cannot track individual particles, the study focused on their aggregate properties. Unlike Dayhoff, whose work was routinely questioned by her peers in biology, Goad had the benefit of working at Los Alamos, with unconditional institutional support. He did not have to justify the validity or importance of his work, as his institutional colleagues at Los Alamos shared in the common premises of his studies.

Los Alamos created a special unit with an Arnold Schwarzenegger sounding name: the "T10" or Theoretical Biology/Biophysics. Their work required intensive calculations, which as Dayhoff's work which were fed into the era's most advanced computers, in this case the IMB 7094.

Yet we might ask why the project was begun in the first place. There was a strange confluence of factors, some of which we have previously seen. The era of the personal computer began in the 1980s. As equally important was the rise of SQL databases, or relational databases which had first been proposed by Edgard Codd at IBM. In short, these can be defined as ways of arranging data whereby the content of the data was separated from its relational structure. One of the benefits of this digital format was that one could change the entire structure of data itself countless of times, looking for patterns, without altering the data itself.

Prior databases had consisted of 'flat tables' 'databases', due to the antiquated mindset of the NIH.

One of the early conditions imposed by the NIH in its funding was that any and all database had to be readable by both men and computers, hence the predominant flat databases.[295] Again, the NIH basically viewed the procedure as simply the digital compilation of previously researched material. It goes without saying that such a presumption implied that the database would not add any particular value to the previously gathered data; the value resided on the original collection of such data. Another reigning presumption of the NIH at the time was that there was a direct proportionality of gene to protein, implying that one could simply track a gene's protein function on a table—a notion which was quickly proven to be incorrect. Again, the NIH continually refused to understand the revolutionary implications of computational biology.

When *GenBank* was first set up by Goad, he followed the NIH requirements, but early on he became aware of it significant limitations. While the contract stipulated that there could only be a three month lag for new databases, by 1986 *GenBank* already had a 10 month lag that was all too rapidly increasing. The key problem was, again, the nature of the flat database, whereby all prior data points had to be continually readjusted after changes in the formal data structure had been enacted. Modifying each and every prior bit of data was a sheer waste of time, which further increased along with the size of its contents—a compounding vicious cycle.

Goad was able to quickly identify the inherent limitations of the system, and began porting all the databases from flat structures to relational ones, a task which was finished by 1989. This simple change allowed the database to grow as quickly as new data was continually being added to it. Its success made it a key precursor to the notion of a human genomic array, later known as the *Human Genome Project* (HGP), by showing that such a project could be financially and technically feasible. Any future changes in the relational structure of such a vast magnitude

---

[295] A flat file might consisted of a but of ATGC, followed by a separator '>'.

could be easily applied long after the sequencing had been performed.

By 1987 there had been a handoff of *GenBank*. One of the preoccupations in the biological community was that Goad had also contracted a non academic institution (BBN) for the distribution of the final material—which in spite of his scientific sagacity was not the wisest of decisions. Too much collegiality in academia leads to uncritical decision making and susceptible to horrible outcomes. Goad's decision raised many questions, in particular that to which is referred to as *cream skimming* in the world of telecommunications. BBN would be getting the best part of process: the final raw data, paying a minimal percent of its original cost. While most of the data had been produced by individual scientists, a private company now stood to profit via its distribution charges. The very scientist which had attacked Dayhoff for her scientific contributions potentially had driven the entire new discipline into the ground.

In 1987, the program was transferred back into the NIH and placed under control by Ostell, now in charge of the *National Center for Biotechnical Information* (NCBI). Ostell continued to revolutionize the field. He turned all the relational information into binary format, specifically ASN (akin to a type of http that was readable by all program), and arranged to place it into a framework of frameworks, also using ASN as its foundation. The brilliance of this move is that it allowed for universal searches across multiple databases, which by then numbered 1,500—or a vast collections of ATGCs. A number of other changes were implemented which continued to open up the scope and range of genetic queries.[296]

Two broad models emerged: a democratic model and a top-down model. In the first, the database essentially amounts to an index of information spread across a whole host of database in different institutions. The database does not contain the information itself, but rather tags where knowledge can be found.

---

[296] The comprehensiveness of such searches led some to the odd claim that bioinformatics was more 'theoretical', claiming that generalization naturally 'flowed' from the data itself without any human cognitive intervention.

One of the benefits of this scheme is its scalability; since it is not responsible for storing and maintaining the original information, it can grow as quickly as the information increases across a whole host of other databases. It does not impose any structure other than common standards for the sharing of information. Another benefit is that the costs of such a system are relatively low, as it does not require a powerful computing structure for its efficient functioning.

The opposite is true of the top-down centralized system, as that found the *Ensembl* database run by the *European Bioinformatics Institute*. It consists of a computer cluster of 549 servers, that while not large by CERN standards, still represents a substantial investment in computing power. While the original information is not produced on site, all information is formally stored in the database. One of the early debates in its design had been whether ontological structures would be imposed on the data, as that proposed by Michael Ashburner 1998, when the fly, mouse, and yeast database scientists got together to establish common standards.

There can be no doubt hat "biology in silica" represents an enormous leap in the world of biology; but a technological euphoria should not overlook the many costs incurred by such a radical change in science.

As in big physics, it is pretty clear that 'biology in silica' will create hierarchical divisions in biology, between the haves and the have nots, due to the high costs involved. If a single machine costs hundreds of thousands of dollars, the costs of a regular laboratory will increase exponentially, various times those of what a traditional 'wet laboratory' would require. Yet, presuming that the funds are successfully raised, 'biology in silica' also introduces problematic trends in the culture of biology— specifically an intellectual bifurcation between those who master advanced statistical stochastic methods and those who do not.

Biology in silica also implies that the field will become a science of 0.01% nations, given that LDC nations will lack the necessary capital required to invest in such a costly science. LDCs will not be able to compete, resulting in social hierarchies which extend across national, educational and personal levels.

Places as Puerto Rico with severe financial shortfalls and a stumbling economy will be barred from entry into the field.

The most noxious tendency, however, might be the resultant changes produced in the character of science. Will original scientific questions be asked, where insight, work, and intuition pay off as in the science of Newton or Einstein? Or will it result in an alleged "hypothesis free" biology where computers do all of the cognitive work for us? Will there be a substitution of the scientist by the computer?

Such is what might occur in the field of bioinformatics, where the promises for a career in science might only be granted to a few and the vast majority of scientists will be left out becoming technicians as voiced in Walter Gilbert's concerns.

Conclusion

Tragically, the recent news for September 2018 provide, yet again, supplementary evidence reinforcing many of the points discussed in the book—in particular the role of information in a society's moral order as predicted by sociobiology.

Two journalists, Wa Lone and Kyaw Soe Oo, who had uncovered the genocidal massacre of the Muslim Rohingya ethnic group in Myanmar, were arrested and placed in a trial which resulted in jail sentences of seven years—terms which are far longer than that which any of the solider perpetrators obtained for committing the actual atrocities. *Reporters without Borders* noted that the two were being arrested for simply doing what reporters do, journalism; worst still, the reporters were framed by the government and subjected to what was essentially a sham trial. It is clear that, as in Unit 731, the government was bent on hiding the facts of the genocide. More shockingly still was Nobel Peace Prize laureate (1991) Aung San Suu Kyi's implicit acceptance of the deeds as Myanmar's primer minister. More that 700,000 Rohingya have fled the nation, after their homes were burnt, their women raped, and their men systematically murdered by the Burmese military forces. Wa Lone and Kyaw Soe Oo were accused of accessing confidential papers, discovering an atrocity recognized by the very military officers at the sham trial.

The same might be said for other cases, as Edward Snowden's revelations in 2013 regarding the extent and range of *National Security Agency* (NSA) communications interception— of a degree and sophistication which, prior to his revelations, would have seemed to be drawn from a science fiction movie rather than a textbook of international affairs. In spite of the constitutional violations such spying entailed, the NSA had been silently monitoring its domestic population for decades; so advanced was their spying that few were aware of its range and depth. The naiveté with which US citizens regarded security in the digital world came crashing to the ground following Snowden's revelations. Paradoxically, this deep inspection by the NSA did not prevent the abuses of digital technologies by other powerful state actors as North Korea or Russia, and the loss of countless millions of dollars and valuable patents and innovations stored presumably in a safe servers. In short, the NSA not only failed to fulfill its military function in the digital arena, but used its capability to violate constitutional limitations of executive power.

Snowden's NSA case is particularly suggestive of the repressive manner in which science, medicine and technology are being used by postmodern developed countries. New sciences and technologies alter the lived reality, and to a degree constitute a kind of implicit 'magic' no different from the wagon wheels which so fascinated colonial Africans. The fictionalized dreams of comic sergeant Dick Tracey is now a common lived reality of countless digital smart watch owners. The traits of this scientific technological repression are, again, framed within the boundaries set by sociobiology. Governments have a particularly sharp allergic reaction to the publicity of its wrongdoing.

Without a doubt, new scientific discoveries and technological innovations increasingly provide far reaching powers that usually stand outside the realm of common experience, undetectable and untraceable when used. When US State Department personnel in Cuba were subject to odd sounds in 2017, these were not mere paranoid fictions of the imagination, but targeted microwaves attacks from a moving van or truck which produced palpable brain damage. This is perhaps no different from the countless number of leaders, journalists or morally-minded citizens that have been subject to some form or other of technological abuse by the state, internal or external. The medical bioterrorism perpetrated by Russia, such as the poisoning of former KGB agents with radioactive compounds, are perhaps the most obvious—but only because its victims were already aware of the mechanism used by the very dictatorial governments they had once been a part of. Some of Salvador Allende's closest colleagues in Chile were assassinated in a similar fashion, and made to appear as natural health-related deaths. It is certainly not known the degree and scale to which such tools have been inflicted upon ordinary civilians in the United States and its territories, with no awareness of their existence or nature.[297] If Unit 731 stands as an example, we may suggest that the frequency of their use is much higher than typically presumed.

Since everyone fears 'the long shadow of the future' (Axelrod), there is a high likelihood that any such perpetrators will seek to repress any and all forms of information when it comes to related to personal immoral and unethical acts. A raped girl will sooner be assassinated by her rapists than given her freedom, for sheer threat of future exposition of the perpetrator that she represents; only the fact that pedophilic Catholic priests could instill fear on their young victims prevented grosser abuses from occurring.[298]

In light of the preceding cases, one might be tempted towards a position supporting an extreme form of information transparency, whereby all walls, literal and metaphorical, are torn down. While this will undoubtedly prevent many gross abuses and crimes from occurring, it will also, in and of itself, have a profound social cost: the system of scientific and technological innovation itself.

Innovation and progress require privacy to exist. The personal exchange of ideas result in innovations and discoveries from which the participants, and the community, directly benefit. If we were to establish a wholly transparent community, scientists and innovators stand the risk of being preempted by rivals who have easily intercepted their communications, either stealing good ideas or innovative patent designs—as has occurred in China for many years. Under such a system, the innovation system would grind to a halt, as there would be no incentives for innovators in the form of personal recognition or financial reward.

---

[297] The use of such invisible passive/aggressive weapons contributes to a destruction of a collectivity's moral social order.

[298] One of the striking features of Catholic pedophilia is the amount of time under which such atrocious activities were kept hidden from public sight—including that of the victim's own parents.

Similarly, the role of gossip, once historically beneficial, can easily devolve into detrimental vicious cycles in modern densely crowded urban areas. In sparsely populated areas the smallest bit of information might be useful for survival—the location of a watering hole, for example—the form of communication we know as gossip played a positive survival role in premodern societies. Indigenous Eskimo groups living in frozen areas provide many examples; 'gossip' became a key survival tool for Lewis and Clark in their journey throughout the American West. States as Texas still retain some of these dynamics, in that the geographic distance typically result in a relatively isolated context of activity. However, more densely populated urban areas have transformed the evolutionary environment of behavior and survival: public opinion.[299] As the cost of personally acquired information is high, gossip pertaining to nearby dangers becomes not only an effective 'survival too', but also a weapon in social competition; the greater the competition, the greater its use as an offensive weapon.

What we may then conclude? Is there is little likelihood that a consistent public policy can be established with regard to information transparency and its impact on the moral social order? The briefly reviewed extremes of information transparency both have their associated set of costs and benefits, suggesting that the careful balancing of forces in any given instance is the best policy to follow. But this is not an acceptable answer, which provide no guidelines towards policy.

To provide one, we have to return back to the origins of ethics and place these firmly within the framework of sociobiological theory.

It is clear that the Lockean *tabula rasa* model had serious limitations, particularly with regard to its social applications. While institutions do have a certain amount of leverage in altering human behavior, there are clear limitations as to their

---

[299] While the rapid acquisition of information is incurred at a low personal cost, the likelihood of its abuse increases dramatically. Even academics who are careful in their scrutiny of printed information fall into the easiest of errors by coming to rely upon networked sources of information, which is ultimately not verified.

impact. Humans are not like silly putty, who can be transformed at will by the state. Catholic moral codes are also based on obsolete notions of human nature and have to be substantially modified. Rather, we may invert the issue, and notice that the conclusions of sociobiological theory are more in concordance with Rousseau's notions.

Men, as all primates, are born ethical, but it is within a social context where the expression of their internal moral voice becomes altered and distorted. The masks men wear distort the external expression of their internal moral voice. While humans have a built in tendency to be good and ethical, evolutionary theory implies that their behaviors are not inflexible, but rather vary according to circumstance.[300] Morality, as with all other biological structures of animal morphology, varies according to the environment, defined both the natural and the social surroundings.[301]

One consequence of the flexible nature of human morality is that in order to establish the most dignified and harmonious societies possible, we have to critically understand how underlying public institutions, in whatever shape or form, result in unethical behavior, broadly defined. As noted by Herbert Spencer, while men have the intention of doing good, if they are not given the opportunity to express their beneficence, they will not. Oppressive social conditions in the long run can undermine the most determined of good souls, just as easily as it can squash the most intelligent and able of a community, as shown by eugenics. We may point to the iniquities of global financial transactions as an example, specifically the variations in the price of national currencies.

Enormous variations in the relative value of different national currencies leads to a rise in prostitution in nations with low relative valuations. As is typical in national politics, while the transaction might be 'rational' from a participant's point of view, it is detrimental at the macrosocial level. Proceeding from two

---

[300] If it were otherwise, they would not be able to dynamically adapt to new conditions, and human species as a whole would not exist.
[301] In this, the Catholic notion that a moral society can only be obtained via sermonic moral strictures is a grossly fallacious claim.

different genetic pools, a young fertile woman can obtain a month's worth of national salary from a single encounter, while a man who would not otherwise have the possibility of such reproductive opportunity obtains it. At the macrosocial level, however, the structural currency differentiation establishes distorted social roles which become structurally fixed in socio-economic relations; its problematic nature was well described by George Simmel more than a century ago.

Contrary to prior British idealizations, extreme Malthusian competition inevitably degrades human behavior. Overpopulation and excessive competition over limited resources pits men against each other in a Hobbsian *bellum omnium contra omnes*, and draws out the worst features of humanity. The best cases of this dynamic are areas of extreme poverty, ghettos, *favelas*, and *caserios*, where the grossest forms of dehumanization exist. This, in turn, implies that for however much the FBI might attempt to control crime and drug activity in a Puerto Rican *caserio*, until the more fundamental issue of Malthusian overpopulation is resolved, the effort over the long run will be futile.[302] Bringing greater political awareness to crime, as suggested by Carlos Pabon, will also fail to resolve the issue for similar reasons.[303]

A social policy which ends up reproducing Malthusian competition in toto is an incredibly foolish policy, at best. Tragically, policies designed by liberal historians, conservative pundits, and the Catholic church tends towards the same ends, albeit with differing mechanisms.[304] The rejection of population control, for whatever reasons, will degenerate social conditions over the long run, increase crime and reduce human dignity. As

---

[302] Other historical cases as British urban areas during early industrialization provide further evidence of this dynamic as well.

[303] Its true cause has not been publicly identified.

[304] While liberals criticize population control as a form of colonial eugenics and conservative pundit pretend that abortion—the removal of undeveloped cells—is a form of assassination, their policies do nothing but serve to increase the incidence of crime in the communities they wish to assist. Similarly, while the Catholic Church wishes to foment 'peace and good will amongst men', their continual attack upon practices such as abortion do nothing but result in a world opposite to that of their stated goals.

well noted by Steven Levitt and Stephen Dubner in *Freakonomics* (2011), *Roe v. Wade* was a landmark decision which resulted in a precipitous drop in crime during the 1990s. Pope Francis's courageous attempt to reform the Catholic Church goes a long way to readjust its policies towards grater concordance with current scientific knowledge—a process which the institution, qua institution, has historically been hesitant to undertake. It is unclear whether Pope Francis will be successful in his insightful and farseeing ambition, and unfortunately recent evidence suggests that he will not.[305]

Yet overpopulation is not the only social condition whereby man is reduced to a beast. In their attempt to 'do harm' to criminals, prisons in fact help to perpetuate the criminal activity it is meant to suppress. If we abide by sociobiology's findings, the curative path for a prisoner, help him recover his lost morality, is perhaps the most durable route over the long run. Whether or not the reader agrees with this stance, there can be no doubt that the dehumanization of man does little to improve his presumptions regarding the nature of society and the character of social relations in a community. If society becomes his 'enemy', then by definition he will act according to these parameters.

A greater number of analyses have to be undertaken with regard to many modern institutions, as implied by the innovative work of Morris (*The Human Zoo*). Just as many political leaders do not quite grasp the destructive power of modern nuclear weapons (Robert J. Art) or the complexity of pollution (Price), the impact of institutions in which actors exist is poorly understood. The US Founding Fathers were well aware of their influence, given their detest of political parties, which grossly degraded the rational conduct of men.

The vast amounts of wealth produced by the modern corporation places the human brain in a social context very different from that of its evolutionary origins, and as its institutional counterpart (prison) also degrade men's souls. The

---

[305] Pope Francis was recently attacked for having been aware of pedophile activity of its officials dating to 2000. There can be no doubt that this attack is a pretext by the conservative faction of the church to get rid of such a liberal theologian.

corporate fortunes give the appearance of unlimited resources, a situation the exact opposite of actual human evolutionary setting, and hence the personal notions of a complete freedom of action—regardless of its effect upon the rest of the citizenry. Only their enormous financial power and resources prevent more corporate leaders from being arrested for criminal wrong-doing—though not entirely so. Nations as El Salvador and South Korea are taking bold steps in ensuring that political and corporate leaders stand trial and criminal sentencing for wrongdoing; their example should be more abundantly imitated.

With regard to the role of ethics in biology, these will tend to be highly politicized, and many scholars have unfortunately committed errors along the way. The well known historian of science Anita Guerrini has been correctly attacked by animal rights groups for providing a biased historical rationalization for animal experimentation. While the latter has been undeniably useful in science, there are many other routes to knowledge and, as Harvey's case tells us, animal experimentation did not provide definitive proofs of scientific truth. Guerrini has used her reputation to carry the argument for the institutionalization of such experimentation.[306] However, writers on the left, as the Jesuit priest Jose Ferrer, on the other hand err by turning ethics into a strictly philosophical issue, as if ethical dilemmas were wholly distinct from the historical periods in which they occur.[307]

Sadly, their discussion of bioethics have been historically disingenuous as well—a development which tragically destroyed the institutional origins of history of science in Puerto Rico. While such postures were driven by a predominant concern with the abuse of genetics in the modern world, a return to medieval scholasticism will certainly not 'fix' any-and-all ethical debates in the field of biology.

---

[306] Guerrini had been a doctoral student of Richard Westfall, the leading scholar of Isaac Newton after dedicating 30 years to the topic.

[307] This contradiction is implicit in their redefinition of bioethics, situating ethics in biology within a particular historical period—a philosophical approach Giambattista Vico was also forced to employ within the Catholic strictures of the seventeenth century.

Yet the principal problem with Ferrer's bioethics is that, at its core, such philosophical stances deny evolution and the evolutionary antecedents of human behavior. Contrary to Catholic presumptions, human nature cannot be freely molded by any institution, no matter how powerful it might be, but are rather bound within particular cognitive structures and delimitations. While we today reject teleological notions imbuing man with godly traits, countless experiments have shown that a sense of justice permeated our earliest primate ancestors—and hence, it has to be noted that our sense of justice is fortunately something that cannot be easily removed or modified. An individual may or may not listen and act according to their own inner voice, but this does not mean that the inner voice has disappeared.

Epilogue

After writing the book, I came across the excellent works of Jill Lepore and Timothy Snyder, which reflect many of the key themes from this study. While Snyder describes the rise of authoritarian regimes via the deterioration of public information, Lepore describes how the modern redefinition of progress excluded its moral aspects during twentieth century United States. Regrettably, the repressive use of medicine in modern semi-peripheral colonial regions as Puerto Rico, so clearly demonstrated in the cases of Cornelius Roads and Pedro Albizu Campos, continues to this day.

—R

Bibliography

Acosta y Calvo, Jose Julian de, ed., *Historia Geográfica, Civil y Natural de la Isla de San Juan Bautista de Puerto Rico de Iñigo Abbad y Lasierra*. Madrid, España: Doce Calles, 2011.

Advisory Committee on Human Radiation Experiments, *The Human Radiation Experiments*. New York: Oxford University Press, 1996.

Andreas-Holger Maehle. *Drugs on Trial: Experimental Pharmacology and the Therapeutic Innovation in the Eighteenth Century*. Amsterdam: Rodolpi, 1999.

Angell, Marcia *The Truth about the Drug Companies: How They Deceive Us and What to do About It*. New York: Random House, 2004.

Annas, George J. and Michael A. Grodin. *The Nazi Doctors and the Nuremberg Code: Human Rights in Human Experimentation.* New York: Oxford Univesity Press, 1992.

Aponte-Vázquez, Pedro *The Unsolved Case of Dr. Cornelius P. Rhoads: An Indictment*. San Juan, Puerto Rico: Pedro Aponte Vazquez, 2004.

Ardent, Hannah. *The Origins of Totalitarism*. New York: Harcourt, Brace and Jovanovich, 1973.

Art, Robert J. and Robert Jervis, *International politics: enduring concepts and contemporary issues.* Boston, MA: Perasons Press, 2017.

Artigas, Mariano Thomas F. Glick, and Rafael A. Martínez, *Negotiating Darwin: The Vatican Confronts Evolution, 1877-1902.* Baltimore, MD: Johns Hopkins University Press, 2006.

Ashford Bailey K. *A Soldier in Science: The Autobiography of Bailey K. Ashford*. New York: William Morrow and Co,

1934.

Assimov, Isaac. *Breve Historia de la Biología*, trad. Ricardo Zelarayán. Buenos Aires Argentina: Editorial Universitaria de Buenos Aires, 1966.

Ayala, Francisco J. "The Biological Foundations of Ethics" *Revista Portuguesa de Filosofia*, 66,3 (2010), pp. 523-537.

Berlin, Isaiah. *Three Critics of the Enlightenment: Vico, Hamann, Herder, Henry Hardy* ed. Princeton, New Jersey: Princeton University Press, 2000.

Bliss, Michael. *The Discovery of Insulin*. Chicago, IL: University of Chicago Press, 1982.

Brosnan, Sarah F. & Frans B. M. de Waal, "Monkeys reject unequal pay" *Nature* 425 (18 sept. 2003), 297-299.

Campbell, Joseph. *The Hero with a Thosuand Faces*. Princeton, NJ: Princeton University Press, 1973.

Capshew, James H. and Karen A. Rader, "Big Science: Price to Present" *OSIRIS*, 2nd series, 7 (1992), 3-25.

Catlin George. *North American Indians: Letters and notes on the manners customs and Conditions*, Vol 1. London, UK: Dover, 1973.

Champion Michael and Lara O'Sullivan, eds, *Cultural Perceptions of Violence in the Hellenistic World*. New York: Routledge, 2017.

Chow-White, Peter A. and Miguel García-Sancho, "Bidirectional Shaping and Spaces of Convergence: Interactions between
Biology and Computing from the First DNA Sequencers to Global Genome Databases," *Science, Technology, & Human Values*, 37, 1 (January 2012), pp. 124-164.

Cohen, I. Bernard. *Benjamin Franklin's Science*. Cambridge, MA:
Harvard University Press 1990.

-----. *Science and the Founding Fathers: Science in the Political Thought of Jefferson, Franklin, Adams, and Madison*. New York, New York: W. W. Norton & Co. , 1995.

Coleman, William. *Georges Cuvier, Zoologist: A Study in the History of Evolution Theory*. Cambridge, MA: Harvard

University Press, 1964.

Congressman Edward Markey Report, *American Nuclear Guinea Pigs: Three Decades of Radiation Experiments on U.S. Citizens*, 99th Congress 2d Session, 65-0190. Washington DC: US GPO, 1986, HTML: http://nsarchive.gwu.edu/radiation/dir/mstreet/commeet/meet1/brief1/br1n.txt.

Conrad, Goeffrey W. and Arthur A. Demarest, *Religion and Empire: The dynamics of Aztec and Inca expansion.* Cambridge: Cambridge University Press, 1984.

Corning, Peter A. "The Science of Human Nature and the Social Contract" *Cosmos and History: The Journal of Natural and Social Philosophy* 11, 1 (2015), 15-40.

Cosans, Christopher E. "Galen's Critique of Rationalist and Empiricist Anatomy," *Journal of the History of Biology*, 30, 1 (Spring 1997), pp. 35-54.

Darwin, Charles. *The Descent of Man*. New York: Prometheus Books, 1998.

Dawkins, Richard. *The Selfish Gene*. New York: Oxford Univesity Press, 1976.

Degler, Carl N. *In Search of Human Nature: The Decline and Revival of Darwinism in American Social Thought*. New York: Oxford University Pres, 1991.

Denis, Nelson A. *Guerra Contra todos los Puertorriqueños: Revolución y Terror en la Colonia Americana*. New York: Nation Books, 2015.

Dennett, Daniel "Animal Consciousness: What Matters and Why" *Social Research* 62,3 (Fall 1995), 691-710.

Diamond, Jared. "Race without Color," *Discover* 115,11 (Nov. 1992), 31-37.

Dietrich, Alexa S. *The Drug Company Next Door: Pollution, Jobs, and Community Health in Puerto Rico*. New York: New York University Press, 2013.

Duany, Jorge. *Puerto Rican Nation on the Move: Identities on the Island and in the United States*. Durham, University of North Carolina Press, 2002.

Duprey, Marlene. *Bioislas: Ensayos sobre biopolítica y gubernamentalidad en Puerto Rico*. San Juan, Puerto

Rico:
Ediciones Callejon, 2010.

Durant, Will *The Life of Greece, The Story of Civilization*, Vol 2. New York: Simon and Schuster, 1939.

Ehlrich, Paul R. *Human Natures: Genes, Cultures, and the Human Prospect*. New York: Penguin Books, 2002.

Faber,Paul Lawrence. *The Temptations of Evolutionary Ethics.* Berkeley, CA: University of California Press, 1994.

Fernandez Lynch, Holly, "Ethical Evasion or Happenstance and Hubris? The US Public Health Service STD Inoculation Study" *Hastings Center Report* 42, 2 (Mar-Apr 2012), 30- 38.

Fernós, Rodrigo. *From Galieo to Boltzmann: A History of the Fragility and Resilience of Science*. Corpus Christi, TX: VirtualBookworm, 2016.

Ferrer, Jorge Jose. *Deber y Deliberación: Una invitación a la bioética*. Mayaguez, Puerto Rico: Centro de Publicaciones Académicas, 2007

-----. . "La bioética como quehacer filosófico" *Acta Bioethica* 15,1
(2009), pp. 35-41.

Forbes, Nancy *Imitation of Life: How Biology is Inspiring Computing*. Cambridge, MA: MIT, 2005.

Frankenburg, Frances Rachel. *Human medical experimentation : from smallpox vaccines to secret government programs*. Santa Barbara, California : Greenwood, 2017.

Franklin, Benjamin. *Autobiography*. New York: Library of America, 2005.

Freeman, Derek. *Dilthey's Dream: Essays on human nature and culture*. Acton, Australia: Austrlian National University Press, 2017.

French, Roger *Ancient Natural History: Histories of Nature*. London, UK: Routledge, 2004.

Freud, Sigmund. *Civilization and Its Discontents*. New York: W. W. Norton, 1989.

Gabriel, Mordecai L. and Seymour Fogel, *Great Experiments in Biology.* Engelwood Cliffs, NJ: Prentice-Hall Inc., 1955.

Gambetta, Diego. *Codes of the Underworld: How criminals*

communicate. Princeton, NJ: Princeton University Press, 2009.

Geison, Gerald L. *The Private Science of Louis Pasteur*. Princeton, NJ: Princeton University Press, 1995.

Gilmore, David G. *Manhood in the Making: Cultural Concepts of Masculinity.* New Haven: Yale University Press, 1990.

Glick,Thomas F. "Science and Independence in Latin America (with Special Reference to New Granada)", *Hispanic American Historical Review*, 71, 2 (May, 1991), pp. 307-334.

Gold, Hal. *Unit 731 - Testimony*. New York: Yenbooks, 1996.

Goldsmith, Timorthy H. The *Biologica Roots of Human Nature: Forging Links Between Evolution and Behavior*. New York: Oxford University Press, 1991.

Goliszek, Andrew. *In the name of science : a history of secret programs, medical research, and human experimentation.* New York : St. Martin's Press, 2011.

González Ávila, Manuel. "Exploraciones sobre las conexiones de la ciencia con la ética y la política. Discusión sobre las influencias y acciones recíprocas de la ciencia con los procesos sociopolíticos en el marco general de la ética" *Revista Umbral* 6 (Diciembre 2011), pp. 84-106.

Goozner, Merrill *The $800 Million Pill: The Truth Behind the Cost of New Drugs* Berkeley, CA; University of California Press, 2004.

Gould, Stephen Jay. *The Panda's Thumb: More Reflections in Natural History*. New York, New York: WW Norton & Co., 1980.

Guerrini, Anita *The Courtiers' Anatomists: Animals and Humans in Louis XIV's Paris* Chicago IL: University of Chicago Press, 2015.

-----. *Experimenting with Humans and Animals: From Galen to Animal Rights* Baltimore, MD: Johns Hopkins University Press, 2003.

Hardy, G. H. *A mathematician's apology*. London, UK: Cambridge University Press, 2009.

Harris, Steven James. "Jesuit ideology and Jesuit science: Scientific activity in the Society of Jesus, 1540-1773",

Ph.D. Thesis The University of Wisconsin –Madison, 1988.

Headrick, Daniel. *Tools of Empire: Technology and European Imperialism in the 19th Century.* New York: Oxford University Press, 1981.

Henkin, David M. *The Postal Age: The Emergence of Modern Communications in Ninteenth-Century America.* Chicago: University of Chicago Press, 2006.

Hernandez Marrero, Damián. "La moral y la no Posibilidad de la ética como ciencia en Ludwig Wittgenstein" *Diálogos*, 97 (2015), pp. 107-119.

Herrnstein Richard J. and Charles Murray, *The Bell Curve: Intelligence and Class Structure in American Life.* New York: Free Press Co., 1994.

Hippocratic Oath

Hofstader, Richard. *Anti-Intellectualism in American Life* (New York: Vintage Books, 1962).

-----.*Social Darwinism in American Social Thought* (NY: George Grazilles Inc., 1955).

Johnson, Linda, " Animal Experimentation in 18th-Century Art: Joseph Wright of Derby: An Experiment on a Bird in an Air pump," *Journal of Animal Ethics* 6 , 2 (2016), pp. 164–176

Kevles, Daniel J. *In the Name of Eugenics: Genetics and the Uses of Human Heredity.* Berkeley, CA: Unviersity of California Press, 1985.

Kitcher, Phillip, *Science in a Democratic Society.* New York: Prometheus Books, 2011.

Korom, Frank J. "The Evolutionary Thought of Aurobindo Ghose and Teilhard de Chardin", *Journal of South Asian Literature* 24,1 (Spring 1989), 124-140.

LaFleur, William R. and Gernot Böhme, eds. *Dark Medicine: Rationalizing Unethical Medical Research.* Bloomington, IN: Indiana University Press, 2008.

Lavoisier, Antoine and Pierre Laplace, "Memoir on Heat' by (1780) in Gabriel, Mordecai L. and Seymour Fogel, eds, *Great Experiments in Biology.* Engelwood Cliffs, NJ:

Prentice-Hall Inc., 1955.

Layton, Edwin. "Mirror-Image Twins: The Communities of Science and Technology in 19th Century America," *Technology and Culture* 12 (1971): 562–80.

Lederer, Susan E. *Subjected to Science: Human Experimentation in America Before the Second World War*. Baltimore, MD: Johns Hopkins University Press, 1995.

Levitt, Steven D. and Stephen J. Dubner, *Freakonomis*. New York: Harper Collins, 2011.

Lewis, Sinclair. *Arrowsmith*. New York: Signet 2008.

Livingstone Smith, David, *The Most Dangerous Animal: Human Nature and the Origins of War.* New York: St. Martin's Press, 2007.

Lloyd, G.E.R. *Aristotle: The Growth & Structure of his Thought*. New York: Cambridge University Press 1990.

Losco, Joseph "From outrage to orthodoxy? Sociobiology and political science at 35," *Politics and the Life Sciences* 30, 1 (Spring 2011), pp. 80-84.

Maehle, Andreas-Holger. *Drugs on Trial: Experimental Pharmacology and the Therapeutic Innovation in the Eighteenth Century*. Amsterdam: Rodolpi, 1999.

----- "The Ethical Discourse on Animal Experimentation, 1650-1900", in Andrew Wear; Johanna Geyer-Kordesch; Roger K. French, eds., *Doctors and ethics : the earlier historical setting of professional* ethics. Amsterdam ; Atlanta, GA : Rodopi, 1993 pp 203-251.

Mayr, Ernst. *Toward a New Philosophy of Biology: Observations of an Evolutionist.* Cambridge, MA: Belknap Press, 1988.

McDonald, Forrest. *Novus Ordo Seclorum: The Intellectual Origins of the Constitution* Lawrence, KS: University Press of Kansas, 1985.

Mendez, Elizabeth "The Study that Helped Spur the U.S. Stop Smoking Movmeent" *American Cancer Society*, Jan 9, 2014, HTML: https://www.cancer.org

Merton, Robert K. "A note on science and democracy" *Journal of legal and political sociology* I (Oct. 1942), 116-26.

Michael, Bliss. The Discovery of Insulin. Chicago, ILL:

University of Chicago Press, 1982.

Milford, Jessica "Experiments Behind Bars: Doctors, drugs and prisoners", *Atlantic Monthly* (Jan 1973), pp. 64-73.

Mircea, Eliade. *Essential Sacred Writings from Around the World.* New York: Harper Books, 1987.

-----. *The Sacred and the Profane: The Nature of Religion.* San Diego, CA: Harcourt Brace Jovanovich, 1987.

Morange, Michel. "The Death of Molecular Biology?" *History and Philosophy of the Life Sciences*, 30, 1 (2008), pp. 31-42.

Morris, Desmond. *The Naked Ape* (New York: Random House, 1967).

-----. *The Human Zoo.* New York: McGraw-Hill Boo, 1969.

Moyers, Bill, ed. *The Power of Myth by Joseph Campbell.* New York: Turtleback Books, 2012.

Nanjundiah, Vidyanand "Role of Mathematics in Biology," *Economic and Political Weekly*, 38, 35 (Aug. 30 - Sep. 5, 2003), pp. 3671- 3677.

Oliver Vázquez, Marlen Ed., *Ensayos en bioética: una perspectiva puertorriqueña.* San Juan, Puerto Rico: UPR, Recinto de Ciencias Médicas, 2013.

Pabón Ortega, Carlos. *Polémicas: política, intelectuales, violencia* (San Juan, Puerto Rico: Ediciones Callejon, 2014).

Petryna, Adriana. "The Competitive Logic of Global Clinical Trials" *Social Research*, 78, No. 3 (Fall 2011), pp. 949-974.

-----. *When experiments travel : clinical trials and the global search for human subjects.* Princeton : Princeton University Press, 2009.

Petryna, Adriana and Andrew Lakoff, Arthur Kleinman, Eds. *Global pharmaceuticals : ethics, markets, practices.* Durham. NC: Duke University Press, 2007.

Pinker, Steven *The Blank Slate: The Modern Denial of Human Nature.* New York: Viking, 2002.

Post, Stephen G. *Encyclopedia of Bioethics*, 3rd Edition. New York: Macmillan Reference USA, 2004.

Presidential Commission for the Study of Bioethical Issues, *Ethically Impossible: STD Research in Guatemala from 1946-1948*. Washington DC: US GPO, 2011. Html: www.bioethics.org.

Price, Don K. *America's Unwritten Constitution: Science, Religion, and Political Responsibility.* Baton Rouge: Louisiana State University Press, 1983.

-----. *The scientific estate.* Cambridge, MA: Belknap Press, 1965.

Prieto, Andrés Ignacio. "Los naturalistas jesuítas: Naturaleza, evangelización y propaganda en Sudamérica, 1588-1676." PhD Thesis, University of Connecticut, 2006.

Rachels, James *Created from Animals: The Moral Implications of Darwinism.* New York: Oxford University Press, 1990.

Reverby, Susan M., "Will the STI studies in Guatemala be remembered, and for what?" *Sexually Transmitted Infections* 89, 4 (Jun 2013), 301-3).

-----, "Normal Exposure and Inoculation Syphils: a PHS 'Tuskegee' Doctor in Guatemala, 1946-1948" *Journal of Policy History* 23, 1 (2011), 6-28.

-----, "Ethical Failures and History Lessons: The U.S. Public Health Service Research Studies in Tuskegee and Guatemala" *Public Health Reviews* 34,1 (2012), pp. 1-19

Richards, Robert J. *Darwin and the Emergence of Evolutionary Theories of Mind and Behavior.* Chicago, IL: University of Chicago Press, 1987.

Ridley, Matt *The Origins of Virtue: Human Instincts and the Evolution of Cooperation* New York: Penguin Books, 1998.

Rivera Ortiz, Ángel Israel. "La ética pública en la educación sobre
ciencia política, gobierno y administración pública" *Revista*
de *Administración Pública* 42 (2011),
p1-32.

Rodriguez, Angel Ricardo "Infectious Imperialism: Race, Syphilis, and Human Experimentation in Guatemala City, 1946-1948", MA Thesis, University of California, Santa Barbara, 2014.

Rose, Steven ed., *The Richness of Life: The Essential Stephen Jay Gould*. New York: W. W. Norton, 2006.

Ross, Colin A. *The C.I.A. doctors : human rights violations by American psychiatrists*. Richardson, TX : Manitou Communications Inc., 2006.

Ruiz Marrero, Carmelo. *El Gran Juego de Ajedrez Botánico: Escritos sobre biotecnología y agroecología, 1999-2014*. San Juan, Puerto Rico: Editorial Tiempo Muerto, 2015.

-----, "Biotechnology in Puerto Rico: Myths and Hazards", *Synthesis/Regeneration* 43 (Spring 2007), 33-35.

-----, "Frankencrops in the Caribbean" *Synthesis/Regeneration* 52 (Spring 2010), 23-24.

-----, "More GM Crops in Puerto Rico," *Synthesis/Regeneration* 58 (Spring 2012), 2-5.

Ryan, Kenneth J. "Research Misconduct in Clinical Research: The American Experience and Response, " *Acta Oncologica* 38,1 (1999), pp. 93-97.

Sarasohn, Judy *Science on Trial: The Whistle-blower, the Accused, and the Nobel Laureate*. New York: St. Martin's Press, 1993.

Santos y Vargas, Leonides, "Bioética y Sociedad" *Puerto Rico Health Sciences Journal* 17,1 (March 1998), pp. 155-157.

Simmel, Georg *The Philosophy of Money*, transl. Tom Bottomore and David Frisby London: Routledge & Kegan Paul, 1978.

Simon, Stacy "Study: 50 years fo Anti-Smoking Effort have saved 8 Million Lives" *American Cancer Society*, Jan 7, 2014, HTML: https://www.cancer.org/latest-news/study-50-years-of-anti-smo . . . =2.61867311.2060078296.1526165025-659248635.1526165025

Skloot, Rebecca *The Immortal Life of Henrietta Lacks*. New York: Broadway Books, 2011.

Smith, John Maynard. "The Concept of Information in Biology" *Philosophy of Science*, 67, 2 (Jun., 2000), pp. 177-194.

Smith, Roger *The Norton History of The Human Sciences*. NY: W. W Norton & Co., 1997.

Stepan, Nancy Leys *"The Hour of Eugenics": Race, Gender, and*

Nation in Latin America. Ithaca: Cornell University Press, 1991.

Stevens, Hallam *Life Out of Sequence: A Data-Driven History of Bioinformatics* Chicago, IL: Unviersity of Chicago Press, 2013.

Talmor, Ezra *Mind and Political Concepts* (Oxfrod, UK: Pergamon Press, 1979).

Turner, Fred. *From Counterculture to Cyberculture: Steward Brand, the Whole Earth Network and the Rise of Digital Utopianism* Chicago: University of Chicago Press, 2006.

Udias, Agustin *Jesuit Contribution to Science: A History.* New York: Springer, 2015.

Walsh, John P. and Todd Bayman, "Computer Networks and Scientific," *Social Studies of Science*, 26, 3 (Aug., 1996), pp. 661-703.

Washington, Harriet A. *Medical Apartheid: The Dark History of Medical Experimentation on Black Americans from Colonial Times to the Present*. New York: Anchor Books, 2006.

Welsome, Eileen. *The Plutonium Files: America's Secret Medical Experiments in the Cold War*. New York: Delta Pub., 2000.

Wright, Robert, *The Moral Animal: Evolutionary Psychology and Everyday Life*. New York: Vintage Books, 1994.

Wright, Susan. "Recombinant DNA Technology and Its Social Transformation, 1972- 1982" *Osiris,* Vol 2 (1986), pp. 303-360.

Wrigley, E. A. *Population and History*. New York: McGraw-Hill Book Co.,1969.

Zimmerman Umble, Diane. *Holding the line : the telephone in Old*

Order Mennonite and Amish life. Baltimore : Johns Hopkins University Press, 1996.

Index

Rodrigo Fernós

334

www.ingramcontent.com/pod-product-compliance
Lightning Source LLC
Chambersburg PA
CBHW071533200326
41519CB00021BB/6473